Pathogenic root-infecting fungi

PATHOGENIC
ROOT-INFECTING
FUNGI

S. D. GARRETT, Sc.D., F.R.S.

Fellow of Magdalene College, Cambridge, and
University Reader in Mycology

CAMBRIDGE
AT THE UNIVERSITY PRESS, 1970

Published by the Syndics of the Cambridge University Press
Bentley House, 200 Euston Road, London N.W.1
American Branch: 32 East 57th Street, New York, N.Y.10022

© Cambridge University Press 1970

Library of Congress Catalogue Card Number: 72–10024

Standard Book Number: 521 07786 9

Printed in Great Britain
at the University Printing House, Cambridge
(Brooke Crutchley, University Printer)

Contents

Preface

This book is a sequel to my *Biology of Root-infecting Fungi*, published by the Cambridge University Press in 1956 and reprinted in 1960. Since the appearance of that book, the subject has advanced so rapidly that it has been easier to write an entirely new book than to attempt a second edition of the first one. I have been particularly concerned to keep this book within the length of its predecessor, so that it can be read through rather than used merely for reference; in this, at least, I have been successful, though I must point out that comprehension of any chapter depends upon a general knowledge of what has gone before it. I have made the necessary economies in two ways. First, I have reduced the space given to ecology of the soil fungus flora, because this is now better known and understood, and so less in need of discussion here. Secondly, I have omitted a chapter on non-pathogenic, i.e. mycorrhizal, root-infecting fungi as my title indicates, though I have referred to work on these fungi where appropriate. For this I have to thank my friend Professor J. L. Harley at the University of Oxford, whose most welcome monograph on *The Biology of Mycorrhiza* first appeared in 1959 and has made a chapter on these fungi in my new book quite superfluous.

No plant pathologist needs to be reminded that pathogenic root-infecting fungi constitute a substantial and omnipresent threat to the welfare of all our crop plants. Somewhat neglected during the earlier years of plant pathology, largely owing to the technical difficulties of their investigation, they are now receiving their due share of attention from plant pathologists and soil microbiologists. I need stress no further this economic aspect of their importance to biologists, except to remark that more information, more cooperation and more financial assistance are available for the study of an organism of economic importance than for one that is not; all university biologists must now be aware of this, though some may still refuse to act upon it.

An interest in the applied aspects of biology is nowadays more likely to be accompanied by an equal interest in its more fundamental aspects, and vice versa, because the two sides of the subject have become increasingly interdependent. And so I hope this book may interest some biologists other than mycologists and plant pathologists, because my subject has opened for me, and for others too, a window upon a world wider than that of the diseased plant. This world is that of the living soil, which is the mother of all terrestrial plant life and therefore of direct concern to every human being and of interest to every biologist. Inside the soil, many kinds of living organism are interacting continuously with one another and with the root systems of green plants, and so diseases of the root system cannot be studied in isolation as a simple interaction between root system and fungal pathogen; much more than this is involved, and so the root-disease investigator must be concerned with everything that goes on in the soil. As a plant pathologist he is often perplexed and baffled by the intricacies of his problems, but as a biologist he is surprised and delighted by the insights he thus gains into the complex, microcosmic world of the living soil. And so I hope in this book the zest of the biologist will compensate for the worries of the plant pathologist, whose work is not finished when he has elucidated the biology of his disease; he must then attempt to translate this fundamental knowledge into successful practice, and this second problem is often more intractable than the first.

Here is perhaps the best place to mention a matter of taxonomic importance to mycologists, and that is the nomenclature practice I have tried consistently to follow. With a very few exceptions, I have referred to every fungus by the taxonomic binomial for its perfect (sexual) state (where one is known), giving the name of the imperfect (asexual) state in parentheses, at least at the first mention of the species. I have given names of authorities for the binomial of the perfect state at the first mention only of each species, but this information can quickly be found by reference to the general subject index. A few fungi, such as *Rhizoctonia solani* Kühn, are more commonly known by the names of their imperfect states, especially when taxonomists disagree about the disposition of the perfect state; in such cases I have followed the common usage. Wherever possible, I have followed the nomenclatural practice recommended by the late W. C. Moore in his *British Parasitic*

Fungi, published by the Cambridge University Press in 1959; this has saved me much time and trouble. Lastly, I have used the scheme of Snyder & Hansen (1945) for the classification of species and formae speciales of the genus *Fusarium*, which contains many important species of root-infecting pathogen.

In conclusion, I wish to thank all those who have helped me in various ways, directly or indirectly, with the preparation of this book. First my colleague Dr John Rishbeth, for his generosity in reading and commenting upon the whole of my typescript. This is but part of a much larger debt I owe to him, because I have gained many valuable insights into our common field of research through the work that he and his associates have carried out over the last twenty years and more, and from our frequent discussions. Secondly, I wish to thank Mr W. J. Bean, chief technical assistant in our Subdepartment of Mycology, for his invaluable help over many years with the whole of our research programme, and in particular for the amount of time he has saved for me personally. Thirdly, I am grateful to all those authors and editors who have kindly allowed me to reproduce their illustrations, which will add much to the value of my text; sources of these have been acknowledged individually as they occur. Lastly, I thank Miss Ruth Braverman for the great care and skill with which she typed from my manuscript.

S. D. G.

1 Introduction

The root and shoot systems of flowering plants inhabit two quite different environments, the soil and the subaerial environment, respectively. Both environments together affect the health and vigour of the plant; sometimes together but more often separately, they affect the activity, survival and dispersal of pathogenic fungi. Because these two environments are so different, root-infecting fungi constitute a natural ecological group of pathogens that are distinct from air-borne fungi causing diseases of the shoot system. In either case, a thorough understanding of the plant's partial environments is a prerequisite for effective investigation of a disease and so a brief comparison between these two environments will be the best introduction to the special features of root-disease investigation.

In the soil, the buffering effect of the soil mass does not permit the wide fluctuation in temperature and atmospheric humidity that often characterizes the subaerial environment of the plant's shoot system; throughout much of the year over most of the earth's cultivated surface, the soil is both warm enough and moist enough for some microbial activity to go on. Sources of energy for this continual microbial activity are provided both by the dead remains of plants and animals and by excretion of organic nutrients by the root systems of living plants and the living bodies of animals. Thus in every soil at all times when temperature and moisture content permit, the soil microflora of fungi, bacteria and actinomycetes is active in the decomposition of organic substrates, from which it derives energy and essential nutrients for the synthesis of microbial protoplasm. Also active in the soil when physical conditions permit is the soil fauna, which is divided according to size range into micro-, meso- and macro-fauna. The micro- and meso-fauna occupy existing soil spaces, whereas members of the macro-fauna, such as the earthworm and the mole, make new soil spaces for themselves by burrowing. The soil metabolism is accelerated by members of the soil fauna, through their comminution, ingestion and digestion of organic material; in natural, uncultivated soil, it is members of the macro-fauna, and above all the earthworm, that effect incorporation of surface litter with the soil mass. Pathogenic organisms, whether plant or animal in their biological classification, are thus living in an environment of often intense biological activity. Root-infecting pathogens that are classified in the Plant as distinct from the Animal Kingdom belong to the following divisions: Fungi, Bacteria, and Actinomycetes. In addition to these, some plant-infecting viruses are transmitted from plant to plant through introduction into the root system by various fungal pathogens (Grogan & Campbell, 1966). Pathogenic root-infecting fungi, however, much exceed all other groups of root pathogens both in number of species and in economic importance, and it is with these root-infecting fungi that this book will be concerned.

Pathogenic root-infecting fungi thus live for the most part in an environment fundamentally different from that occupied by air-borne fungal pathogens infecting the plant's shoot system. For root-infecting fungi, competition with and interference from saprophytic soil micro-organisms is generally a more powerful limiting factor than is restriction by the physical conditions of the soil environment; as we have already noted, most cultivated soils at most times are both warm enough and moist enough for microbial activity, including root infection by pathogens, though the pace of this activity is determined by these physical factors. In contrast, both dispersal of and infection by air-borne fungal pathogens of the shoot system are critically dependent upon physical factors of the subaerial environment, such as temperature, wind velocity, atmospheric turbulence, humidity and incidence of rainfall (Gregory, 1961). Infection of the shoot system is affected by temperature and critically limited by the need either for a film of liquid water or for a high relative humidity that is essential for spore germination on the leaf or other surface; the powdery mildews (Erysiphaceae), alone amongst air-borne fungal pathogens, have so evolved as to escape this restriction.

THE ROOT SYSTEM OF FLOWERING PLANTS

Flowering plants vary almost as widely in the morphology or growth habit of their root systems as they do in their shoot systems. As pathologists, however, there is one distinction that fundamentally concerns us, and that is the one between angiospermic monocotyledons, which have no secondary thickening, and dicotyledons, in the perennial species of which the tap root and main lateral roots increase indefinitely in thickness through the production by the cambium of secondary wood (xylem) and bark (phloem). This continued activity of the cambium is paralleled by that of the cork cambium (phellogen), as a result of which the root becomes encased in an increasingly thick layer of cork tissue. The older, cork-covered parts of the root system in trees thus offer a substantial mechanical barrier to invasion by root-infecting fungi; in addition, such large roots can muster a more powerful active resistance to attempted fungal invasion than can the more slender roots of monocotyledons and herbaceous dicotyledons. As we shall see later, this highly evolved degree of resistance to fungal infection possessed by tree root systems has resulted in a parallel evolution of infection habit amongst root-infecting fungi parasitic upon them.

The root-systems of many forest trees are also peculiar in another way, in that the ultimate rootlets are regularly mycorrhizal. In such ectotrophic mycorrhizas, the root hairs of the uninfected feeding rootlet are replaced by a sheath of fungal tissue, from which a mantle of hyphae sometimes extends outwards into the soil. Much evidence has now accumulated for the view that this ectotrophic mycorrhizal association characteristic of forest trees is a true symbiosis. The host plant provides the fungus with its carbon substrate in the form of sugars; the mantle of fungal hyphae can absorb nitrogen, phosphate, potash and other soil nutrients more efficiently than can the root hairs of an uninfected rootlet, and this difference is maximal in infertile soils. Thus for both partners in this association, symbiosis provides more efficiently and more securely than independent nutrition can do. Detailed evidence in support of this conclusion has been discussed by Harley (1969) in his recent monograph entitled 'The Biology of Mycorrhiza'.

Problems concerning nutritional exchange between flowering plants and this and other types of mycorrhizal fungi were earlier epitomized by Harley (1968) in his Presidential Address to the British Mycological Society.

HOST RESISTANCE TO INFECTION

Every plant pathologist, whatever the disease he may be studying, is fundamentally concerned with the operation of host resistance to infection. This book is not the place in which to consider in any depth or detail so large and fundamental a part of the whole science of plant pathology. Fortunately such a survey has recently been published by Wood (1967) in his admirably comprehensive book *Physiological Plant Pathology*. Mechanisms of disease resistance in the root system do not differ in any fundamental way from those that operate in the shoot system, and so I shall not discuss in any detail the way in which these mechanisms are thought to operate. But I shall be much concerned with the way in which variations in host resistance affect the behaviour of root-infecting fungi, and for this reason I have provided here a schematic classification of types of host resistance and disease escape to serve as a convenient reminder of the wide variety of these mechanisms.

This schema is intended to serve as a general guide rather than as a rigid classification. In particular, as Wood (1967) has noted, it is difficult to distinguish between some types of pre-existing resistance due to fungistatic substances already present in the healthy cell, and infection-provoked resistance; often a difference between these two postulated mechanisms may be merely one of degree, depending upon the speed and readiness with which a non-toxic precursor present in the uninfected cell is transformed into a fungistatic substance. Secondly, functional resistance due to the opening/closing behaviour of stomata and flowers, which I have included in this schema for the sake of completeness, does not seem to have any parallel in the behaviour of the root system. One of the best known examples of such functional resistance is provided by the greater susceptibility of rye than of wheat to the ergot disease (*Claviceps* spp.); this is due not to greater tissue

DISEASE RESISTANCE (1, 2)
(±absolute *or* environmentally conditioned)

(1) PRE-EXISTING RESISTANCE (*a*, *b*, *c*)

(*a*) *Mechanical*	(*b*) *Functional*	(*c*) *Chemical*
Cuticle, cork, thickened and lignified cell walls, etc.	Opening/closing behaviour of stomata, flowers, etc.	Acid cell sap, phenolics, glycosides, alkaloids, etc.

(2) PROVOKED RESISTANCE (*a*, *b*, *c*)

(*a*) *Production of fungistatic substances*, e.g. phytoalexins, resin, gum, etc.	(*b*) *Morphogenetic responses of individual cells to infection*, e.g. thickening and alteration of cell walls, lignitubers around invading hyphae, tylosis of xylem vessels by associated parenchyma cells, etc.	(*c*) *Wound meristems*, producing cork barriers

DISEASE ESCAPE (1, 2)

(1) *Pathogen absent*	(2) *Pathogen present, but environmental conditions inhibit infection*

susceptibility of rye to infection, but simply to the fact that flowers of rye remain open for much longer periods than do those of wheat. Functional resistance of this type has often been classified as 'disease escape' but I prefer to regard it as a type of disease resistance, because it is an intrinsic, genetically determined characteristic of the species or variety of plant.

If we define disease resistance as including all characteristics of the species or variety that are intrinsic and genetically determined, then under the category of 'disease escape' we are left with those situations in which a genetically susceptible host escapes infection either because the pathogen is not there or because, though present, it is inhibited from infecting the host by unfavourable environmental conditions. To say that a plant escapes a disease because the pathogen is absent is seemingly to invite that derision lying in wait for all pedagogues, but it so happens that escape from some root diseases operates in a way that is not at all obvious at first sight. To take an example from my own experience, I found that susceptibility of wheat roots to infection by the take-all fungus

5

(*Ophiobolus graminis* (Sacc.) Sacc.) was increased by a high or unbalanced plane of nitrogen nutrition (Garrett, 1941). Yet at that time it was already well known that field resistance of wheat crops to take-all could be greatly enhanced by generous application of nitrogen or of any other major soil nutrient, most commonly phosphate, that happened to be in short supply. The solution to this seeming paradox lay in recognition of the fact that if one or more major nutrients is limiting growth, then this deficiency curtails the power of the wheat plant to produce crown roots up to its maximum potential capacity (Garrett, 1948). The new roots produced by a well-nourished plant are not individually more resistant to infection than are those produced by a poorly nourished plant; indeed, if the nitrogen content of the soil solution is unduly high, individual roots may be more rather than less susceptible, as I have already remarked. But if random infection over any area results in 50 % of the crown roots becoming severely infected, then a well nourished plant producing 60 roots will be left with 30 functioning roots, whereas a poorly nourished plant producing 20 roots will be left with only 10 functional ones. The well nourished plant can thus produce fairly well filled ears in spite of losing half of its functional roots through disease, but the poorly nourished plant is likely to produce virtually no harvestable grain. This relationship between the 'shoot load' and the number of functional roots required for its effective support, so that a full complement of grain can be ripened, has been studied by White (1947). Such a difference caused by crop nutrition is not an example of environmentally conditioned resistance; it is clearly a case of disease escape.

It is further worth noting that seedlings of most plant species must owe their survival largely to escape from damping-off diseases and seedling blights caused by *Rhizoctonia solani* Kühn (perfect state = *Thanatephorus cucumeris* (Frank) Donk), *Pythium* and *Fusarium* species and similar pathogens. Young seedlings have only a low resistance to infection by these unspecialized parasites, which are mostly widespread saprophytes in cultivated soils. But though these pathogens are widespread, their distribution in the soil is local around the sites of past substrates, and so

6

survival of seedlings is due to disease escape through local absence of a pathogen, much more commonly than to possession of any significant degree of resistance to infection.

Declining resistance of senescent root systems

The consequences of a significant decline in resistance of the root system, or of any part of it, on the behaviour of a specialized root-infecting fungus already established at one or more points on the root system, and perhaps spreading along it, will now be considered. Such a decline in root resistance to infection is likely to follow any occurrence that reduces vigour of the plant as a whole. Amongst natural hazards of this kind can be listed attack of the shoot system by diseases and pests, and damage by grazing animals, by fires of spontaneous origin (e.g. lightning strike), by high winds, floods and other natural agencies. To these can be added such agricultural and sylvicultural hazards as deliberate pasturing of stock, spraying crops with inadequately selective weed killers, and felling of trees.

The first consequence of such a decline in root resistance following upon a decline in general plant vigour is that specialized root-infecting fungi begin to spread much more rapidly over the root system as soon as the limiting factor of host resistance has been removed, or at least abated. Similarly, any still viable fungus that has been localized by host resistance in a stationary lesion is now likely to break out and resume active spread. R. Leach (1939) was the first plant pathologist to bring this important effect to general attention. He showed that the felling of forest trees in East Africa, in preparation for clearing land for tea plantations, induced a spectacular increase in rate of spread of *Armillaria mellea* (Fr.) Kummer over the (stump) root systems, and that this constituted a most serious disease hazard for the young tea bushes that were subsequently planted on the cleared forest site. Similarly, an early practice by some Australian farmers of grazing down young winter wheat crops in order to consolidate the land, and hence to delay underground spread of the take-all fungus, was sometimes found to increase the incidence of take-all at harvest;

7

in such instances, the beneficial effects of soil consolidation must have been outweighed by the adverse reduction in plant vigour. I myself have seen a similar result in increased incidence of take-all to follow the spraying of wheat crops with some of the earlier weed killers.

A sufficient decline in plant vigour, and therefore in resistance of the root system to infection, thus accelerates infection and spread by the specialized parasites. A further decline in the level of root resistance does not necessarily operate to their advantage, because it permits invasion first by weak parasites and then by obligate saprophytes. Host resistance thus affects the specialized parasites in several ways; at its maximum, it restricts their infection and spread but at an intermediate level it is their chief protection against invasion of their living substrate by weak parasites and saprophytes, which come in during the final phase of dwindling tissue resistance.

An exploitation of weakly parasitic and saprophytic fungi for controlling the spread of *Armillaria mellea* on the stump root systems of forest trees in East Africa was most ingeniously devised by Leach (1937) through his ring-barking control method. He found that ring-barking the forest trees a year or more before felling resulted in a much accelerated death of the root system. In such ring-barked trees, the roots are still kept at work absorbing water and nutrient salts from the soil in order to satisfy the demands of the transpiring crown of foliage, but are denied the supply of carbohydrates that in intact trees passes down the bark of the trunk from the photosynthesizing leaves. The root systems of such ring-barked but standing trees therefore exhaust their starch reserves much more quickly than do stump root systems, which are not performing work in absorption. At the time he published this paper, Leach thought that a high starch content of root tissues *per se* favoured infection by *A. mellea*, and that ring-barking trees made the roots unfavourable for invasion by this fungus because of their reduced starch content. In the light of subsequent work, however, it now seems clear that the operative effect of ring-barking is rapidly to reduce root resistance to so low a level that weakly parasitic fungi and then obligate saprophytes can invade the tissues. This decline in root resistance is indeed closely correlated with a decline in reserve starch, for the simple reason that active tissue resistance to infection is fuelled by carbohydrate reserves and when these are finally exhausted, both life of the root tissue and all active resistance come to an end. This conclusion is consistent with the results of extensive field trials at the Rubber Research Institute of Malaya on tree and stump poisoning with arboricides, at first sodium arsenite and later 2,4,5-trichlorophenoxyacetic acid (2,4,5-T), which were undertaken with a view to hastening invasion of the root system by weak parasites and saprophytes. Considerable success was obtained with these poisoning treatments in restricting development of *Fomes lignosus* (Klotzsch) Bres. and other specialized root parasites on root systems of the old trees and their subsequent transfer to roots of the young rubber trees; this result is compatible with the above conclusion that it is rapid killing of the old root systems rather than a reduction in their starch content that constitutes the mechanism of this particular example of biological control.

INOCULUM POTENTIAL

Consideration of host resistance to infection is logically followed by an attempt to analyse the nature of the *invasive force* of the parasite, which tends to operate in such a way as to overcome host resistance and hence to permit infection by the parasite to proceed. During the course of experiments on the logistics of infection by the rhizomorphs of *Armillaria mellea*, I realized the need for a precise definition of this 'invasive force' of the parasite, for which I adopted the term 'inoculum potential'. As this term was already current in the literature but had been used sometimes in a narrow sense and sometimes in a broad and rather ill-defined sense, I was obliged to redefine it, in order to express the idea of invasive force as precisely as possible. My definition was as follows: *inoculum potential is defined as the energy of growth of a parasite available for infection of a host, at the surface of the host organ to be infected.* To this I added a necessary corollary: 'implicit in this definition of inoculum potential as the energy of growth of a parasite is the

9

qualification of energy of growth *per unit area of host surface*, without which any such definition would be meaningless'.

Such a definition will have little heuristic value unless it is followed by an attempt to analyse, at first theoretically and then experimentally, what factors contribute to maximal energy of growth in the invading parasite. Inferences from existing knowledge will take us some way in the desired direction. First we can assume that, other things being equal, the energy of growth of a fungal parasite will be directly proportional to the cross-sectional area of the fungus in contact with unit area of host surface. Thus a dense suspension of spores applied to the host surface will have a greater total energy of growth than will a dilute suspension. Similarly, several thousand hyphae aggregated close together in the single organ known as a fungal *rhizomorph* will have a greater energy of growth than the more widely separated hyphae in unorganized mycelium. But the density of fungal hyphae per unit area of host surface, whether present as germ tubes from germinating spores, as hyphae in unorganized mycelium or as hyphae aggregated into a rhizomorph, is not the only factor determining energy of growth of the fungus. Of equal importance will be the relative degree of *vigour* of the fungal hyphae attempting to invade the host. Other things being equal, vigour of the fungal hyphae is primarily determined by the nutrient status of the protoplasm in their apical regions. In my own study of the logistics of infection by rhizomorphs of *Armillaria mellea*, I found that speed of infection of potato tubers increased with increase in volume of the woody inoculum block and decreased with increasing distance between inoculum and host surface. This second finding can be interpreted on the assumption that the rate at which nutrients can be mobilized at the growing apex of a rhizomorph decreases with increasing distance between rhizomorph apex and the inoculum food-base (Garrett, 1956*b*).

The term 'energy of growth' clearly implies that infective hyphae of one sort or another are *actually growing* out from the inoculum, i.e. it implies more than the assumption that the inoculum is *capable* of producing hyphae at a certain density per unit area of host surface and having a certain vigour, as determined by the nutrient status of the hyphal tips, once they have started to grow. For this reason, energy of growth from the inoculum is conditioned also by a third factor, the collective effect of environmental conditions. As environmental conditions vary from optimal to completely inhibitory, so they determine the actual or realized energy of growth ranging from a possible maximum (depending on volume and nutrient status of the inoculum) down to zero. It is further important to note that endogenous nutrients of the inoculum may be augmented by exogenous nutrients from the environment and most commonly by exudation of nutrients from the host surface; root exudates are of particular importance to infection by root-infecting fungi.

In view of the variety of meanings with which this term 'inoculum potential' has been employed by earlier workers, R. Baker (1965, p. 419) has performed a useful service by producing a schema in which various concepts of inoculum potential have been related both to one another and to cognate concepts concerning factors controlling disease development, such as host susceptibility/resistance and environmental effects.

Earlier criticisms of this term 'inoculum potential' *sensu* Garrett (1956 *a*, *b*) were concerned with the difficulty of defining 'energy of growth' with adequate precision. Since then, fresh experimental evidence has come to hand that makes both this concept and its usefulness more explicit; biological concepts differ from those of the physical sciences in that the complexity of biological phenomena often imposes a limit on the precision with which definitions can be made. Nevertheless, a definition that would not content a physicist is often good enough for its purpose in explaining one of the grosser phenomena with which biologists are primarily concerned. In defining what determines energy of growth by an inoculum, the most helpful data come from experiments on infectivity of fungal spores, because the range in volume of these infective propagules falls within definable limits for any fungal species, and within still narrower limits for any population that is used in inoculation experiments. Thus early work by F. T. Brooks (1908) and by W. Brown (1922*a*) demonstrated that the infectivity of spore populations of *Botrytis cinerea*

Fr. on leaves of lettuce and broad bean (*Vicia faba*), respectively, was greatly increased when they were sown on the leaves in a nutrient solution (grape juice and turnip juice, respectively) instead of in water. Brown observed that one effect of the exogenous nutrient was to increase the percentage of spores germinating on the leaf surface; from his data, it is impossible to separate this effect of exogenous nutrient in increasing inoculum potential of the spore dose from the other possible effect, i.e. an increase in nutrient status and hence in vigour of growth by the germ tubes of individual spores. But in the data obtained by Last (1960) in experiments on infection of leaves of *Vicia faba* by spores of *Botrytis fabae* Sardiña, any significant effect of nutrient on percentage germination of spores can be excluded. Maximum percentage germination (approx. 93%) of spores in water was maintained as cultures of *B. fabae* from which the spores were taken aged from 25 to 40 days, whereas infectivity for bean leaves declined to one-tenth of the original at 25 days and to one-hundredth at 35 days age of the parent culture. Infectivity of the ageing spores could be restored by suspending them in orange juice which contains, in addition to proteins, minerals and vitamins, approx. 1·5% citric acid and 4·5% sucrose. When tested separately, sucrose at 4·5% was found to be even more effective than orange juice in restoring infectivity and was followed in order of effectiveness by glucose, mannose and maltose. Last has concluded 'Ageing conidia of *B. fabae* seem to contain adequate reserves for germination but insufficient to meet the demands for infection'. Indeed, we should expect that carbon reserves in the germinating spore would limit the energy of growth and the infectivity of the germ tube more commonly than does any other nutrient, because carbon reserves are used up by respiration of the dormant spore, whereas nitrogen and other mineral nutrients are not.

Whereas Last had demonstrated an effect of *exogenous* nutrients on the infectivity of his conidia, Phillips (1965) was able to compare the infectivity of conidia with different levels of *endogenous* nutrients, by producing macroconidia of *Fusarium roseum* f. sp. *cerealis* (Cooke) Snyder & Hansen (synonym *Fusarium culmorum*

(W.G.Sm.) Sacc.) on culture media of high and low nutrient level, respectively. The comparisons were between Czapek's liquid medium at full strength and at 1/10 strength, and between Czapek's medium with normal glucose concentration (20 g/l) and with 1 g glucose/l, respectively. In each case, conidia produced at the higher nutrient level were more infective to carnation cuttings as expressed by a disease index; percentage spore germination of the low-nutrient conidia (87%) was actually a little higher than that of the high-nutrient conidia (78%).

Less precise data on the effect of volume and/or nutrient status on the infectivity of mycelial inocula can be found scattered through the literature of the last 30 years and more. For fungi infecting the roots of herbaceous plants, an inoculum disc approx. 1 cm diameter cut from the growing margin of a fungal colony growing on nutrient agar (with carbon source equivalent to not less than 2% glucose) is usually adequate for infection. Reduction of inoculum volume much below this will result in failure to infect, as Whitney (1954) has demonstrated with *Helicobasidium purpureum* Pat. (stat. mycel. = *Rhizoctonia crocorum* Fr.) on carrot. The inoculum potential for infection of wheat seedlings by *Ophiobolus graminis* growing on filter-paper cellulose and nutrient salts is determined by the nitrogen level of the nutrient solution when this is the factor limiting rate of cellulose hydrolysis to sugar (Garrett, 1967). But for infection of older parts of tree root systems, which are well protected both by an encasing cork layer and by various active defence mechanisms, inocula of nutrient agar do not suffice and only a woody inoculum of sufficient volume is adequate for infection (de Jong, 1933; Altson, 1953).

SYNERGISM VERSUS INDEPENDENT ACTION DURING INFECTION BY FUNGAL SPORES

From the evidence discussed in the foregoing section, there is no doubt that for any host–parasite combination, there is a threshold value for energy of growth from the inoculum below which infection fails to occur. Sometimes this limiting value of growth energy is exceeded by that of a single, typical fungal hypha of the particular species; in other instances, it is not. Fungi infecting older parts of the root system in trees do not infect by means of individual hyphae, but rather as hyphal aggregates, such as mycelial sheets, strands and rhizomorphs. Here we can postulate with reasonable certainty that synergism between individual hyphae is essential for successful infection. No such assumption can be made, however, for fungal spores that are freely dispersed by wind, rain-splash or moving water. Mechanisms for dispersal of these spores operate in such a way as to render it unlikely that more than a few spores will land on the same infection site at the same time, and so we can expect that infectivity of the average spore will be adequate for infection of the average host individual. A difficulty in accepting this statement arises not from any observations on natural occurrence of spore infections, but rather from the results of inoculation experiments with a range of spore doses. Results of some of these infectivity titrations have shown that the ED50 dose of spores (i.e. the dose giving infection of 50% of the test hosts or host organs) is higher, and sometimes much higher, than might have been expected. This discrepancy between observation and expectation has been sufficient to cause some plant pathologists, from Heald (1921) onwards, to postulate a need for synergism between germ tubes of spores in the act of infection. A survey of recent work, however, will show that this difficulty of interpretation is more apparent than real; as so often happens in scientific investigation, theoretical and experimental models have represented inadequately the natural situation.

Animal pathologists have shown more interest in the mechanism of the infective dose than have plant pathologists, until recently. Halvorsen (1935) seems to have been the first to distinguish clearly between the 'hypothesis of synergistic action' and the 'hypothesis of independent action'; the latter postulates that a single infective propagule can establish infection but that the probability of this happening is less than unity. When the ED50 is high, then p, the probability of infection by a single propagule, must be correspondingly low. This is not difficult to understand; in any population of infective propagules, degree of infectivity must vary, like size, around a modal value and so the infectivity of some propagules will be too low for successful infection of any of the host individuals under test. Some propagules, again, will fail to germinate on the host site, and others will fail to infect because of disadvantages of position. In short, we can expect a wastage of infective propagules comparable to the wastage of spores of saprophytic fungi, and of seeds of flowering plants. For animal hosts, an outstanding contribution towards the solution of this problem has been made by Meynell and his associates (Meynell, 1957*a*, *b*; Meynell & Meynell, 1958; Meynell & Stocker, 1957; Armitage, Meynell & Williams, 1965). The most direct test for discrimination between the hypotheses of independent and synergistic action is to make inoculations with single propagules. For bacterial pathogens of animals, this is technically very difficult, because of the small size of the bacterial cell and because the value of p is so low that the number of inoculation trials with single propagules necessary to achieve success makes the attempt quite impracticable. For discriminating between these two hypotheses, animal pathologists have therefore had to use less direct methods, such as the slope of the log-dose/probit-response curve obtained from results of an infectivity titration, comparison of a divided with an undivided dose of propagules, the use of a mixture of distinguishable strains of a pathogen in inoculation, and inferences drawn from a study of latent periods following inoculation with a range of doses (Meynell, 1957*b*; Armitage *et al.* 1965). Most of the evidence presented or reviewed in these papers is more compatible with the hypothesis of independent action than with that of synergistic action.

The evidence available for plant pathogens, mainly leaf-infecting fungi, is at present limited (Garrett, 1960*a*, 1966*a*) but is sufficient at least to suggest explanations for unusually high values of the ED50 dose. First, for specialized parasites the ED50 is usually quite low, as we might expect when a parasite is well adapted to its host. For many years, pure races of rust fungi and other obligate parasites had to be maintained by continuous culture on their living host plants, and purity was established and maintained by single-spore inoculations; success in such inoculations commonly varied from 10 to 50%. For *Botrytis fabae*, the specialized parasite causing chocolate spot disease of field beans, Deverall & Wood (1961) found the ED50 to be less than ten (the lowest spore dose tested) and obtained success in 10% of single-spore inoculations. Wastie (1962) obtained an ED50 value of approx. four spores, and was successful in 13% of single-spore inoculations, which compares well with a value of 16% calculated from his ED50 value. These results can be compared with those from parallel inoculations of bean leaves with *Botrytis cinerea*, in which the ED50 was found to be around 500 spores. This high value can be explained by the observation that *B. cinerea* is not a primary parasite of healthy bean leaves, but is a widespread weak parasite of damaged or senescent leaves and also of floral parts in a wide range of plant species. These results with *B. fabae* and *B. cinerea* thus explain very clearly one reason for high ED50 values; the range in infectivity values of a spore population determines its natural host range, because infectivity of the average spore must be adequate to overcome resistance to infection of the average host individual. In addition to this limitation, infection by spores is usually limited to specific infection sites on the host plant and often further restricted to one period only in the age sequence of a plant organ from youth through maturity to senescence. When spore inoculation trials are made on the wrong host, the wrong infection site or at the wrong time, either complete failure or high ED50 values for infectivity will be the result.

A good example of failure through choosing the wrong infection site is provided by early infection trials with the ascospores of the take-all fungus, *Ophiobolus graminis*. Whereas with leaf-infecting fungi, the leaf-surface flora of bacteria and fungi is usually too sparse (except in the wet tropics or after nutrient exudation following insect attack) to interfere much with leaf infection by spores of fungal pathogens, root-infecting fungi meet with a more substantial interference by the root-surface and rhizosphere microfloras living on root-exudate nutrients. Thus neither I (Garrett, 1939) nor later D. H. Brooks (1965) were able to infect roots of wheat seedlings in natural soil with ascospores of *O. graminis*, even with heavy spore doses, though they could be readily infected in completely sterile soil or sand. We ascribed this failure to

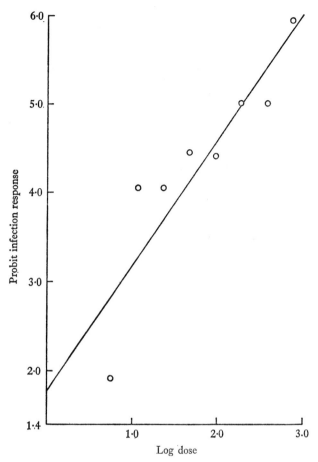

1. Relationship between log-dose of ascospores of *Ophiobolus graminis* and probit infection-response of wheat seedling roots. Drawn from data of D. H. Brooks (1965).

microbial competition for root-exudate nutrients, which appeared to be necessary for successful germination and root infection by the ascospores. By a logical inference from this conclusion, Brooks tried inoculating the proximal region of roots produced by wheat seeds germinating *on the surface of* damp soil, and was immediately successful in obtaining infection; further experiments demonstrated that such spore inoculations were successful if carried out *before* the root surface between seed and soil had become fully colonized by the root-surface microflora growing up the root from the soil below.

Fig. 1 shows a log-dose/probit-response relationship drawn from Brooks's data from an infectivity titration with ascospores of *O. graminis*. The slope of the regression line of y on x is 1·74; this is compatible with the hypothesis of independent action in infection, for which the slope of the line cannot significantly exceed a value of 2·0, otherwise synergism between spores must be occurring. The ED50 value from this infectivity titration was 169 spores, and p, the probability of infection by a single ascospore, was 0·0041. The fairly high ED50 value, and the correspondingly low value of p, can here be ascribed to competition for root-exudate nutrients by the root-surface microflora, which was not absent though certainly at a lower population level than that around roots completely invested by soil. In the latter situation, the ED50 value must have been well above the highest spore concentration tested (8×10^4 ascospores/ml).

Earlier in this discussion, I suggested that in the case of fungal spores individually dispersed by wind, rain-splash or moving water, synergism in infection seemed improbable because natural

Table 1. *Relation between spore load of artificially smutted wheat grain and percentage smut in the crop* (*Marquis spring wheat*) *grown from it*

(From Heald, 1921)

Wt. smut (g) per 100 g grain	Calculated no. spores/grain	Smutted plants (%)
0	104	0
0·005	333	0
0·01	542	0
0·1	5043	7
0·25	19687	15
0·5	34937	1
1·0	59229	25
2·0	96958	56
3·0	183375	15

opportunities for synergism on the infection site would be rare. But clearly this argument does not apply to spores that are naturally dispersed in aggregates, as in insect transmission of spores in a viscous slime. Here the infective propagule is not an individual spore but an aggregate of spores; the occurrence of synergism is not merely possible but even probable. This suggests an explanation for the much quoted data obtained by Heald (1921), who inoculated wheat seed with spores of the bunt fungus, *Tilletia caries* (DC.) Tul., using a wide range of spore doses (Table 1).

From inspection of Table 1, it appears that, in spite of marked irregularities in the run of infection figures, the ED_{50} must have been at least 9×10^4 spores, so it is not surprising that Heald should have concluded that synergism was essential for infection. He may prove to have been correct in this surmise, because bunt spores are sticky and tend to be dispersed, whether by contact or by wind, in clumps rather than singly.

BEHAVIOUR OF POPULATIONS OF INOCULA

So far we have been considering the inoculum potential of a single inoculum in contact with the surface of a root, or of any other underground part of a plant, such as a rhizome, rootstock, corm, bulb, tuber, stool-base, etc. Inoculum potential is the summation of characteristics and situation that decides whether an inoculum will be effective or ineffective; an effective inoculum is one that establishes a successful, progressive infection. An infection can, of course, fail to progress after making a good start, because it may encounter gradually increasing host resistance or locally unfavourable soil conditions at some distance away from the original inoculum. If, however, the original inoculum has produced a corpus of freshly infected host tissue of several times its own volume, then failure of the infection to progress further must be ascribed to causes other than a deficiency of the inoculum.

Any population of inocula distributed throughout the soil of a field, or other unit area having the same cropping and cultivation history, will show a distribution of inoculum potential varying about a modal value; the infectivity characteristics of such a population can be assessed by suitable sampling methods. From such a survey, a prediction can be made of the disease risk that will attend the planting of a susceptible crop. Such information has also been used to explain the occurrence of widespread and severe outbreaks of a particular disease on the 'wrong' type of soil. Thus it has been known for many years that the clubroot disease of crucifers, caused by *Plasmodiophora brassicae* Woron., is favoured by slightly acid soils (approx. pH 6) as well as by adequate soil moisture content and aeration; on alkaline soils (approx. pH 8) the disease is usually of very restricted occurrence. This restriction of the disease on alkaline soils is due to the fact that around pH 8 only a very small proportion of the resting spore population germinates and infects the root hairs (Samuel & Garrett, 1945). It remained for Colhoun (1953) to use this information in a masterly interpretation of the occasional severe outbreaks of clubroot that occur on alkaline soils. He showed that such outbreaks were associated with extremely high resting spore populations of *P. brassicae* in the soil, following continuous cropping with susceptible crucifers. If we assume that 90% of a population of resting spores germinates and infects the root hairs in a soil at pH 6 and that 0·009% does so in a soil at pH 8, then a population 10 000 times larger will be required to produce an equally severe outbreak of clubroot at pH 8. Differences of this order of magnitude in resting spore populations commonly occur. This example illustrates that for predicting disease risk, we require information about all factors contributing to inoculum potential, i.e. about soil conditions as well as about size of pathogen population. And seeing that the occurrence of infection is determined by the balance between inoculum potential of the pathogen and resistance of the host, we also need information about the degree of resistance peculiar to the crop and its variety, and how this is likely to be affected by environmental conditions. Here again, Colhoun (1961) has contributed a valuable example; he found that plants were more resistant to clubroot when grown at low light intensities but that this enhanced resistance could be overcome by a sufficient increase in the resting spore population of *P. brassicae*.

Occasional severe outbreaks of other soil-borne diseases on the wrong type of soil also occur. Take-all of cereals is most prevalent on well aerated soils of light texture, especially if also alkaline. The behaviour of this disease can be correlated closely with the effect of soil conditions on the rate at which *Ophiobolus graminis* spreads over the root system from initial foci of infection, which correspond with distribution of infected plant debris in the soil (Garrett, 1936). If these infected residues from a previous diseased crop, or from weed host plants, are distributed very abundantly throughout the soil, then severe take-all can develop on heavy clay soils and on acid soils, which are usually unfavourable to development of this disease. Here again, a high population level of the pathogen can nullify the restrictive effect of adverse soil conditions upon spread of the pathogen.

SPECIALIZATION OF PARASITISM AMONGST ROOT-INFECTING FUNGI

Root-infecting fungi, like any other group of parasites, can be broadly classified into the two categories of specialized and unspecialized; the group of specialized parasites presents obvious opportunities for further subdivision, as in Chs 3 and 4. This distinction is a fundamental one, and justifies a preliminary discussion in this introductory chapter of the differences between the two groups, and of the relative advantages and disadvantages of these two modes of existence for a parasite. In his book *Natural History of Infectious Disease*, Burnet (1953) has epitomized the biologist's approach to the problem of disease in his aphorism about 'the outlook of a biologist, to whom both man and micro-organism are objects of equal interest'. In order to comprehend the nature of the host–parasite relationship to the fullest possible extent, the biologist has to adopt two mental attitudes in turn; as a parasitologist, he has to identify himself with the parasite, and as a pathologist with the parasitized host.

Unspecialized root-infecting fungi appear as such whichever the criterion for specialization that we choose to adopt. If we choose host-specificity, then we can say that such fungi are usually characterized by a wide host range; a wide host range confers the advantages of abundance and wide distribution upon the parasite. The complementary disadvantage that accompanies host catholicity is that no single host–parasite relationship is of more than transient value to the parasite, a state of affairs paralleled in human sexual relationships by the legendary figure of Don Juan. Indeed, such unspecialized root-infecting fungi are typically ephemeral parasites of seedlings and of young rootlets on older parts of the root system. These juvenile tissues have not yet developed mature tissue resistance to infection. Such fungi also invade wounded roots and the root systems of plants that are suffering from severe nutrient deficiencies, and from exposure to soil toxins and other adverse environmental conditions outside the plant's range of tolerance. Mature parts of the root system in plants growing under satisfactory environmental conditions are adequately protected by host resistance against infection by these unspecialized parasites. If we ask why it is that mature tissue resistance can protect the plant against unspecialized parasites though not against specialized ones, the answer seems to be that the unspecialized parasite does not evade the active defence mechanisms of its host; instead, it cannot avoid provoking them to the maximum extent. Such a parasite invades susceptible living tissues just as a saprophyte colonizes a substrate of dead tissue, rapidly and with production of its full battery of enzymes and sometimes of toxins as well; the first host cells to be invaded are killed, and adjacent cells may also be killed in advance of penetration by diffusion of enzymes and toxins from the incipient infection. This violent disturbance of normal host metabolism serves as a 'warning' to outlying tissues, which then respond by the production of phytoalexins and perhaps also of other chemical agents contributing to the active infection-response of the invaded host. Production of phytoalexins seems to be the plant's most effective reply to challenge by these unspecialized parasites; they seem to protect a plant against all parasites but those specialized to itself, and explain why most plants are resistant to most pathogenic fungi, though the neophyte plant pathologist is apt to forget this (Cruickshank, 1963). Many if not most unspecialized root-infecting fungi can probably exist in the soil as competitive saprophytes in the absence of living host tissues; they have not evolved far enough as parasites to have lost those characteristics that confer a high degree of competitive saprophytic ability, such as a high mycelial growth rate allied to rapid germination of resting propagules when stimulated by nutrient diffusion from a potential substrate, a sufficient armament of tissue-decomposing enzymes, production of fungistatic growth products, including antibiotics, and tolerance of those produced by other micro-organisms. Thus we can visualize such root-infecting fungi as plurivorous though often ephemeral parasites on the root systems of a wide range of host plant species, intermittently with an alternative mode of existence as competitive saprophytes.

The most specialized parasites have evolved in such a way as to cause the minimum possible damage to the host plant functioning as a piece of biological machinery existing for the support and protection of the parasite. The distribution of the specialized parasite is determined by that of its host species and by that alone; in every sense, the success of the specialized parasite has become identified with that of its host species. Sufficient evidence has now accumulated about the nature of specialized parasitism for us to understand, at least in a general way, how this may have evolved. There is evidence that such parasites have discarded, in the course of evolution, most of the equipment that is necessary for life as a competitive saprophyte but is incompatible with the establishment of a harmonious host–parasite relationship; this is beneficial to the parasite in direct proportion to the continuance of effective metabolic functioning by the host plant. Thus we can discern a tendency for the tissue-destroying enzymes of the facultative saprophyte, such as pectinases, cellulases, and ligninases, to be discarded and for the production of toxic growth products to be suppressed. By comparison with those of saprophytes, the growth rate of specialized parasites seems to have been much reduced; when successfully isolated and grown in pure culture, the growth rate of near-obligate parasites like the ectotrophic mycorrhizal fungi has been found to be extremely low. This can be interpreted by supposing that a slow growth and invasion of host tissue by the parasite gives time for mutual adjustment between their respective metabolic processes. Moreover, actual invasion of host cell protoplasts by

23

these highly specialized and often 'obligate' parasites does not occur; as Fraymouth (1956) seems to have been the first to demonstrate, with downy mildew fungi (Peronosporales), haustoria from the intercellular hyphae penetrate the host cell but not its protoplast. The boundary membrane of the protoplast (the plasma-lemma) is invaginated but not punctured. By this membrane-mediated separation of host and parasite protoplasts, a dangerous clash of host and parasite physiologies is avoided; as Nutman (1963) has so aptly phrased it, miscegenation between protoplasm of host and parasite is thereby avoided. The most highly evolved host–parasite relationships that are well known to us are un-doubtedly those of the ectotrophic mycorrhizas of forest trees (Harley, 1968, 1969) and the *Rhizobium*-root nodule association in leguminous plants. Mutual adjustment between host and parasite has here reached such a stage of perfection that the symbiosis has become a single though composite biological organization. This has been well brought out by Nutman (1963) in his discussion of nitrogen fixation in the bacterial tissue of leguminous root nodules; the actual fixation of nitrogen is biochemically effected by co-operation between the bacterium and host membranes, rather than by the bacterium alone.

In specialized fungal parasites that are less highly evolved than the symbionts, the host–parasite relationship does not continue indefinitely but is finally terminated by the behaviour of the parasite. This is exemplified by the so-called 'obligate' parasites that it has until recently proved impossible to grow in culture apart from living host tissues; such are the powdery mildews (Erysi-phales), downy mildews (Peronosporales) and rust fungi (Uredin-ales) amongst air-borne fungi and the Plasmodiophorales amongst soil-borne fungi. Here the eventual disruption of the harmonious phase in the host–parasite relationship is not brought about by any gross physical damage caused by the parasite to the host tissue; rather it is a consequence of *economic damage* resulting from diver-sion of host nutrients to the parasite. The development of some fungal parasites of annual and perennial herbaceous plants during the vegetative phase of the host is so restrained that it is usually impossible to decide by casual inspection whether a plant is infected or not; examples are the loose smut disease of wheat and barley, caused by *Ustilago nuda* (Jens.) Rostr., and the choke disease of grasses, caused by *Epichloe typhina* (Fr.) Tul. Such infections of the host during its vegetative phase have been aptly described as 'latent'. In this type of disease, the normal life span of the host is not shortened and the parasite defers its final economic toll upon its host until the approach of flowering, when the normal host inflorescence is parodied by the appearance of a fungal fructification instead. Similar trends in evolution can be discerned in root-infecting fungi. *Phialophora radicicola* Cain was first described from Canada by McKeen (1952) as an apparently avirulent parasite on the root system of maize. The infection habit is ectotrophic (Ch. 4) and resembles closely that of the pathogenic *Ophiobolus graminis*, but the roots, though infected, remain white and healthy in appearance. This type of infection compares closely with the 'latent infection' by *Ustilago nuda* and *Epichloe typhina* during the vegetative phase of their respective hosts, as just described. Scott (1967) has reported the occurrence of this fungus on the roots of cereals and grasses in the region of Cambridge, and has shown that infection of roots by *P. radicicola* exercises some restriction on spread along the same roots by the take-all pathogen, *O. graminis*, suggesting some degree of natural biological control.

Although most of our knowledge about highly evolved host–parasite relationships has been derived from study of the 'obligate' parasites (Brian, 1967), yet students of the root-infecting fungi have contributed towards this understanding, indirectly and even accidentally. Comparative studies of the competitive saprophytic ability of various pathogenic root-infecting fungi, which will be described in Ch. 5, have demonstrated that the specialized parasites are characterized by a low degree of competitive saprophytic ability, and that their saprophytic life is largely restricted to mere survival in dead, infected host tissues. Their life thus alternates between an expanding parasitic phase and a declining saprophytic one. This restriction of the saprophytic phase of these specialized parasites is enforced by the superior competitive saprophytic ability of unspecialized parasites and obligate saprophytes, which constitute the bulk of the soil fungus flora. It thus seemed appropriate to call these specialized root-infecting fungi by the name of 'ecologically obligate parasites' (Garrett, 1950), thus distinguishing them from the 'nutritionally obligate parasites' that could not be grown as saprophytes even in pure culture (Brian, 1967). But with the successful culture of *Puccinia graminis* f. sp. *tritici* Erikss. & Henn. by Williams, Scott & Kuhl (1966), this distinction between ecologically obligate and nutritionally obligate parasites has become still further blurred; as has happened before and will happen again, an apparently sharp demarcation in the natural world has turned out to be an artefact created by man's imperfect understanding.

SURVIVAL OF PATHOGENIC ROOT-INFECTING FUNGI IN THE ABSENCE OF HOST CROPS

The question of how root-disease fungi survive from one host crop to the next is one of great practical importance for disease control, and therefore much research effort has been expended in the discovery of all possible modes of survival, which are listed as follows:

(1) as competitive saprophytes on dead organic substrates;

(2) saprophytic survival on dead tissues of a host crop or of weeds, infected during the parasitic phase;

(3) dormant survival as 'resting' propagules, e.g. sexually produced oospores and other spores, asexually produced chlamydospores, and multicellular sclerotia;

(4) parasitic survival on living roots and other underground parts of weed hosts and 'volunteer' susceptible crop plants;

(5) parasitic survival on living root systems of plants that show no disease symptoms above ground.

The first three modes of survival will be discussed in Chs. 5, 6 and 7, respectively. The fourth mode of survival, on the underground parts of living weed hosts, will be all too familiar to practising plant pathologists; many of us have found it a useful answer with which to parry awkward questions by farmers who have apparently obeyed all the other instructions for disease control but still have a disease problem with which to reproach their advisers. If we accept the usual definition of a weed as 'a plant out of place', then a 'volunteer' plant, which is one self-sown from an earlier crop, or a descendant of the same, can be classed simply as a weed host.

Evidence of the fifth mode of survival, on the living root systems of plants that show no disease symptoms above ground, has been produced more recently as the result of an almost desperate search for another method of survival when none of the first four listed above seemed able to account for an unusual longevity of the pathogen. Thus the formae speciales of *Fusarium oxysporum* Fr. are specialized vascular wilt pathogens with a narrowly restricted host range as *overt* pathogens, and they show other evidence of parasitic specialization as well. Yet some of them have been found to survive in soil for periods of 10 years in the absence of a host crop and sometimes in the apparent absence of weed hosts as well. Panama disease of bananas, caused by *Fusarium oxysporum* Fr. f. sp. *cubense* (E. F. Smith) Snyder & Hansen in the West Indies and Central America has presented a particularly intractable problem of this kind; saprophytic survival in the comparatively soft and quickly decomposing, infected pseudostem seemed unlikely to account for so great a longevity (commonly 7–10 years) of the pathogen under tropical soil conditions, and weed hosts of the appropriate family (Musaceae) are both uncommon and rather noticeable.

A possible explanation was suggested by Armstrong & Armstrong (1948) and by Hendrix & Nielsen (1958), who found a limited infection of the vascular tracts of the roots and lower stem by formae speciales of *Fusarium oxysporum* in hosts other than the ones in which they caused a typical wilt disease, and after which they were named; overt wilt symptoms, however, were rarely produced in the 'wrong' hosts. These findings cannot be explained away as caused by unnaturally high inoculum potential, because they were made also with plants growing in soils naturally infested by the pathogens. It is certainly possible that parasitic survival of these vascular wilt pathogens in the wrong hosts may contribute to their longevity. Nevertheless, interest in this possibility has declined with the fairly recent realization that these pathogens, like

27

some other species of *Fusarium*, owe their exceptional longevity in soil to the efficiency of their chlamydospores as survival propagules (Ch. 7). Other specialized root pathogens can make much more limited and ephemeral root infections on non-host plants; *Ophiobolus graminis* can infect seedling roots of dicotyledons (Müller-Kögler, 1938) and *Plasmodiophora brassicae* can infect root hairs of some non-cruciferous species (Webb, 1949). In making such limited and ephemeral root infections on non-host plants, the specialized parasites are behaving like unspecialized ones, i.e. they are specialized parasites only on their proper hosts. But the general efficacy of crop rotation in controlling these specialized pathogens (Ch. 8), excepting those that produce long-lived dormant propagules, suggests that parasitic survival on root systems of non-hosts is usually of negligible importance to field practice.

BIOLOGICAL CONTROL OF ROOT DISEASES

Rather more than 40 years ago, a paper by Sanford (1926) on potato scab, caused by the actinomycete *Streptomyces scabies* (Thaxt.) Waks. & Henrici, appeared to arouse root disease investigators into a rather sudden realization that other soil microorganisms must exert a natural biological control of root-disease pathogens, and that it might be possible to increase the degree of this control by appropriate crop husbandry measures. It may always remain a mystery why this general realization did not come sooner, though I have earlier attempted an explanation in a brief history of root-disease investigation (Garrett, 1956*a*, Ch. I). Following Sanford, it soon become apparent that other soil microorganisms interfered with the activity and survival of the pathogens by their competition for soil nutrients and for oxygen, by their excretion of growth-inhibiting by-products including antibiotics, by parasitism (especially by soil fungi) and by predation (soil fauna). Since then, many workers have studied this problem and have concentrated their efforts on two main approaches to the practical problem of obtaining biological control: (1) inoculation of soil or of plant surfaces with selected competitive or antagonistic micro-organisms; (2) alteration of soil conditions by appropriate crop husbandry measures so as to enhance natural biological control by the resident soil micro-flora. Results from inoculation of natural soil with selected competitors or antagonists have been uniformly disappointing; soil is 'biologically buffered' by the resident microbial population and artificially introduced populations, whether of resident or alien species, soon dwindle to negligible size, so that the status quo is restored. Inoculation of host surfaces with a protective coating of a competitive or antagonistic micro-organism is theoretically a much sounder procedure, and Rishbeth (1963) produced such a method for biological control of the root disease of pine trees caused by *Fomes annosus* Fr.; this method is now quite widely practised in Great Britain (Ch. 8). Inoculation of seed surfaces with protective micro-organisms has met with some success in biological control, and in theory could become more widely extended, as R. Baker (1968) has noted in his recent review of biological control, which I can commend as a comprehensive, critical and far-sighted discussion of this subject which, as he remarks, has become 'emotionally loaded' for investigators who have staked heavily on its promise for the future. In Ch. 8, I shall discuss in more detail those methods proposed for biological control through alteration of soil conditions by crop husbandry practices.

In April 1963, an International Symposium was held in the University of California at Berkeley on the general subject of 'Factors determining the Behaviour of Plant Pathogens in Soil'. In 1965, the proceedings of this symposium were published by the University of California Press, under the editorship of Kenneth F. Baker and William C. Snyder, with the title 'Ecology of Soil-borne Plant Pathogens. Prelude to Biological Control'. This first truly international meeting of root-disease investigators will doubtless be recognized by future historians as an important landmark in our subject. The final wording of the title for the published symposium volume is highly significant, implying as it does that successful biological control can be achieved only as a result of the most thorough and fundamental investigations into the microbial ecology of the soil.

With this realization thus made explicit, the subject of biological control, in both its fundamental and its applied aspects, has now become assimilated into the general framework of knowledge about root-infecting fungi and other soil-borne pathogens. The soil population of living micro-organisms (and of macro-organisms too) is an integral part of the soil environment and interacts continuously with the fluctuating physical and chemical factors of this environment. The research worker in his laboratory and the farmer on his fields can do nothing to the soil that does not induce some change in the activity and balance of its microbial population. There must therefore be some element of biological control, or of its reverse, in almost all disease-control practices, including the use of fungicides, insecticides and herbicides. Knowledge and practice of biological control is thus inseparable from general knowledge and control of root-infecting fungi and other soil-borne pathogens, and that is why I have not devoted a special chapter in this book to biological control. Our present outlook may lack the pioneering excitement of 40 years ago, but at least it's nearer to reality and therefore closer to eventual success in harnessing knowledge to practice.

2 Unspecialized parasites

These pathogenic fungi cause diseases of two main types, though the first type can sometimes grade imperceptibly into the second: (1) seedling diseases; (2) diseases of older plants due to a progressive killing of rootlets and sometimes of older parts of the root system as well. Seedling diseases are referred to by various names such as damping-off, seedling blight, etc. The term 'damping-off' was derived from the observed association between this type of disease and overwatering of young seedlings. Diseases due to successional killing of rootlets on forest and plantation trees often manifest themselves only by a gradual decline in vigour of the infected tree, and to some of these the name of 'decline disease' has been given.

SEEDLING DISEASES

Age of seedling and susceptibility to infection

Up to a certain critical age, all the tissues of a young seedling are juvenile and lack mature-plant resistance, so that they are susceptible to infection by unspecialized parasites. Depending upon the volume and distribution of inoculum in the soil, seedlings may be killed before emergence from the soil (high inoculum potential) or after emergence (lower inoculum potential). When atmospheric humidity amongst a stand of seedlings is high, due to overcrowding and overwatering, damping-off fungi spread rapidly from hypocotyl to hypocotyl above the soil surface; the better aeration above the soil surface and the unobstructed growth of mycelium enables the fungal pathogen to spread more rapidly than it can through the soil, and so seedlings may be killed by hypocotyl infection in advance of taproot infection. Such damping-off diseases are most commonly caused by species of *Pythium* and by *Rhizoctonia solani*. The time at which seedling tissues (other than root tips) become resistant to further infection by these pathogens depends both upon the species of plant and upon environmental conditions. Chi & Hanson (1962) found that most seedlings of lucerne and red clover became immune to *Pythium de baryanum* Hesse when 5 days old and that few plants more than 2 weeks old became diseased, though occasional plants older than this did become infected. Working with *Rhizoctonia solani*, Bateman & Lumsden (1965) reported that the hypocotyl of *Phaseolus vulgaris* was highly susceptible to infection during the first 2 weeks, moderately resistant during the third week, and resistant thereafter. They suggested that development of resistance was due to conversion of pectin to calcium pectate in the primary cell walls, and they demonstrated a gradually increasing resistance of these primary walls to maceration by polygalacturonase (PG). Successful infection of older bean hypocotyls by another pathogen, *Sclerotium rolfsii* Sacc. (perfect state = *Corticium rolfsii* (Sacc.) Curzi), was ascribed by Bateman & Beer (1965) to production by this fungus of oxalic acid as well as PG in the infected tissues; oxalic acid was thought to assist the action of PG in two ways (1) by reducing pH value of infected tissues from 6 to 4, the latter value being optimum for PG activity (2) by combining chemically with calcium in the calcium pectates of the primary cell walls, thereby facilitating their hydrolysis by PG.

The effect of temperature upon these seedling diseases is complex but predictable. If the total percentage of seedlings killed by a fungal pathogen is divided into its two components of pre- and post-emergence killing, then it is possible to predict the effect of temperature upon the first component from the relationship discovered by L. D. Leach (1947). Working with a number of seedling host species and pathogens, Leach showed that incidence of pre-emergence killing at different temperatures could be predicted from the ratio:

$$\frac{\text{Growth rate of pathogen}}{\text{Velocity of seedling emergence}}.$$

It is important to realize that Leach's ratio does not predict the effect of temperature upon the *total* percentage of seedlings killed, before *and* after emergence. In order to predict the effect of temperature upon post-emergence killing in various host-pathogen combinations, we have to be able to determine the effect of temperature on the velocity of seedling tissue maturation up to the stage at which seedlings become resistant to further infection. This can indeed be done, though with much more difficulty, as the following examples will show. Early work at the University of Wisconsin and later work elsewhere has demonstrated that seedling tissues mature most rapidly within the temperature range most suitable for growth of the older plant; thus tropical and subtropical species can be described as 'high temperature plants' and temperate-zone species as 'low temperature plants'. This relationship was first demonstrated by the work of Dickson (1923) at Wisconsin on the seedling blights of wheat and maize caused by a single pathogen, *Gibberella zeae* (Schw.) Petch (stat. conid. = *Fusarium graminearum* Schwabe). Whereas seedling blight developed most severely in maize grown at low soil temperatures (8–16° C), it affected wheat seedlings most severely at higher soil temperatures (16–28° C). The optimum temperature for development of seedling blight in each cereal was therefore that furthest removed from the optimum temperature for underground development of the host plant. Dickson, Eckerson & Link (1923) were able to correlate these effects of temperature upon seedling resistance to *G. zeae* with its respective effects on speed of maturation of cell walls in the seedling tissues. A similar effect of temperature upon the resistance of older plants to root infection by *Armillaria mellea* (a specialized parasite) is exemplified by the work of Bliss (1946). He found that the optimum temperature range for root-system development in peach, casuarina, pepper tree, apricot and geranium was 10–17° C; the optimum temperature range for disease development was 15–25° C. The optimum temperature range for root-system development in citrus and rose was 17–31° C, and the optimum range for disease was 10–18° C. For each group of plants, therefore, development of infection and disease was least within the temperature range optimal for root-system development.

Epidemics amongst seedling and older plant populations

In order to distinguish between an epidemic and other types of severe disease outbreak, we have to define what we mean by an 'epidemic'. Both epidemiologists and dictionaries have shown a statesman-like discretion in not committing themselves to a precise distinction, but all plant pathologists would agree that an outbreak of potato blight caused by *Phytophthora infestans* (Mont.) de Bary can almost always be given the title of epidemic. An outbreak of potato blight is typically characterized by rapid spread of the disease from a comparatively small number of primary infection foci by means of wind-borne or splash-borne spores. Making use of this idea of *spread* from primary infection foci as being the essential characteristic of an epidemic, we can give a definition as follows: *an epidemic is an outbreak of disease characterized by a high ratio of secondary to primary infections*. For a soil-borne disease, primary infections are defined as coming from soil inocula, and secondary infections as those resulting from centrifugal spread from the primary infections; some pathogens can spread by free mycelial growth through the soil (see below) but the majority are restricted to spread through actual or proximal root contact between an infected plant and its neighbours. If we stick strictly to this definition, then we realize that severe outbreaks of many types of root disease that have been loosely termed 'epidemics' do not really deserve this appelation. Severe outbreaks of a fungal vascular wilt disease in an annual field crop are certainly not epidemics, because infected plants are so slow to become infective to their neighbours that most of the infections manifesting themselves by the appearance of vascular wilt symptoms must be primary ones derived from soil inocula (Ch. 3). A simulated spread of such a vascular wilt disease during the single growing season of an annual crop may indeed be suggested by centrifugally extending patches of wilted plants, but this can be caused by a centripetal gradient of inoculum potential. Thus plants may wilt earliest in the centre of the disease patch, and be followed gradually by those further and further away from them, because inoculum potential is highest in the centre of the patch and declines centrifugally. Such a gradient in inoculum potential may reflect either a gradient in the distribution of inoculum or a gradient in the suitability of soil conditions for infection from the inoculum.

Perennial plantation crops, on the other hand, occupy their planting sites for periods extending from 5 to 50 years or more; both vascular wilt and ectotrophically spreading parasites (Ch. 4) will continue to spread along the crop root systems from planting time onwards, unless checked by effective control measures. The ratio of secondary to primary infections thus inevitably becomes very high during the extended life of a plantation crop, and so such disease outbreaks can unquestionably be defined as true epidemics.

True epidemics in microcosm are also caused by damping-off fungi in populations of seedlings. Although time for spread of an epidemic is limited by duration of the seedling juvenile phase, yet growth rate of the fungal pathogens is high in relation to the distance between seedlings. Ideal conditions for epidemic development are provided by well watered stands of overcrowded seedlings; under the high atmospheric humidity thus engendered, the pathogen spreads from one hypocotyl to another above the soil surface. The most successful damping-off fungi are therefore the fastest growing ones; species of *Pythium* excel amongst fungi in their rate of growth (up to 15 mm/24 h) though *Rhizoctonia solani* grows at only about half this rate. We can infer from Leach's ratio (above) that *Pythium* species cause more pre-emergence killing than does *R. solani*, and this is indeed confirmed by my own experience.

The characteristically high growth rate of these seedling pathogens is of equal importance for their life in the soil as competitive saprophytes. A high growth rate has been shown to be the most important characteristic for competitive success amongst sugar fungi (i.e. non-decomposers of cellulose) that are pioneer colonizers of virgin substrates (Ch. 5). Thus Martinson (1963) found a high correlation ($r = 0.96$) between competitive saprophytic colonization of soil microbiological sampling tubes (Mueller & Durrell, 1957) by *Rhizoctonia solani* and percentage pre-emergence killing of radishes by this fungus in a number of trials at different levels of inoculum potential. Unspecialized parasites of this type appear to be common soil saprophytes; there is nothing incompatible about these two modes of existence. Although such fungi are common and widely distributed soil inhabitants, seedlings survive by disease escape through local absence of a pathogen. Optimal environmental conditions for growth and tissue maturation do not make young seedlings resistant to infection, but they do shorten the time taken by a population of susceptible seedlings to grow into a population of comparatively resistant young plants.

Mode of host invasion by seedling pathogens

An apparently essential condition for root infection by these seedling pathogens is the production of root exudates by the apical region of young roots; host exudates are also equally important for successful infection of seedling hypocotyls. Root exudates contain sugars and amino acids in nutritionally significant concentrations as well as vitamins and a wide variety of other organic substances (Rovira, 1965). It is these root exudates, and particularly their content of carbon and nitrogen nutrients, that are chiefly responsible for the zone of intensified microbial activity around young roots, which was discovered by Hiltner (1904) and named by him the 'rhizosphere'. Organic nutrients in the root exudates directly affect these fungal pathogens in three ways: (1) they stimulate resting fungal propagules to germinate; (2) they induce directed growth (chemotropism) towards the root by fungal hyphae and directed swimming movement (chemotaxis) by zoospores in soil moisture films; (3) as exogenous nutrients, they increase the inoculum potential of the fungal pathogen.

The relative intensity of nutrient exudation from various regions of young roots of avocado was demonstrated most elegantly by Zentmyer (1961) with zoospores of *Phytophthora cinnamomi* Rands. Zoospores accumulated at the root surface more abundantly in the region of root elongation than at the root tip or in the region of differentiation. Dukes & Apple (1961) similarly observed the attraction of tobacco roots to zoospores of *Phytophthora parasitica* Dastur var. *nicotianae*, and by means of capillary tube experiments demonstrated comparable zoospore responses towards solutions of various sugars and of casamino acid. Cunningham & Hagedorn (1962) studied the attraction of young roots of pea and other legumes to zoospores of *Aphanomyces euteiches* Drechsl. and confirmed Zentmyer's observation that the elongating region of the root attracted the zoospores more strongly than did the tip or the region of differentiation; zoospores were not attracted to the root hairs. Royle & Hickman (1964a, b) again confirmed that maximum attraction was exercised by the elongating region of pea roots to zoospores, in this instance to those of *Pythium aphanidermatum* (Edson) Fitzpatr. A wide range of substances likely to be present in root exudates was tested by these authors in capillary tube experiments; only glutamic acid, after weak base adjustment, and mixtures of sugars combined with mixtures of amino acids induced the pattern of zoospore response observed towards living roots.

Some evidence has also accumulated to suggest that composition and concentration of root exudate nutrients may determine the degree of root infection and pathogenesis. In south-west Ontario, *Rhizoctonia fragariae* Husain & McKeen causes much strawberry root degeneration in cold soil but comparatively little in warm soil; this was correlated by Husain & McKeen (1963) with the effect of temperature over the range 5–30° C on the nutrient content of strawberry root exudates as assessed by mycelial growth of *R. fragariae*. The amino acid content of low-temperature root exudates was significantly higher than that of high-temperature exudates. In three varieties of *Phaseolus vulgaris*, Schroth & Cook (1964) found a significant correlation between amount of exudation from germinating seeds and susceptibility of young seedlings to pre-emergence killing by *Rhizoctonia solani* and species of *Pythium*. Rate of sugar exudation by germinating pea seedlings and infection by *Pythium ultimum* Trow were both found by Kerr (1964) to increase with soil moisture content. Since values for soil moisture content and air space are complementary, a rising moisture content is accompanied by decreasing availability of oxygen. Brown & Kennedy (1966) observed that a reduction of atmospheric oxygen to 4% increased pre-emergence killing of soybean seedlings by *Pythium ultimum* and that this was accompanied by an increased rate of exudation of sucrose, fructose, galactose and glucose by the germinating seedlings. From a study of the behaviour of *P. ultimum* in an artificial 'soil' maintained at different moisture regimes, Griffin (1963 *b*) has concluded that the activity of this and other species of *Pythium* in wet soils must be ascribed as much to a high tolerance of poor gas exchange as to a positive need for abundant water (Griffin, 1963 *a*).

The process of seedling infection by *Rhizoctonia solani* is now understood better than that by any other seedling pathogen, thanks to the work published by N. T. Flentje and his associates at the Waite Agricultural Research Institute of the University of Adelaide over the past decade. This work has also led to a better understanding of the process of root infection by the specialized ectotrophic root parasites (Ch. 4). Kerr (1956) first demonstrated by means of a cellophane bag technique that roots of lettuce and radish seedlings stimulated mycelial development of *R. solani* in diffusion exchange with them through the cellophane film, and that the fungus excreted substances toxic to the roots. Kerr & Flentje (1957) found two distinct stages in the infection process: (1) the hyphae of *R. solani* become attached to the cuticle of the seedling hypocotyl above the junction of epidermal cells, and grow along these junction lines; (2) multicellular infection cushions are produced, each from a branch hypha in which normal elongation has ceased but in which prolific side branching now takes place. Strains of *R. solani* not pathogenic to a particular species of seedling failed to become attached to the hypocotyl and to organize infection cushions thereon. When grown on cellophane over exudates from host and non-host species, infection cushions were formed on the films overlying host exudates, but not on those over non-host exudates; on films over glucose-peptone broth etc., the specific morphogenetic response of infection cushions again failed to appear. But although infection cushions were produced on cellophane over host exudates, hyphal penetration of the cellophane did not follow. But it did follow on strips of host-hypocotyl epidermis placed on host exudate solidified with agar; on this, the normal 'infection' process occurred just as on the hypocotyl of a living seedling. But on host-hypocotyl epidermal strips placed over water agar, attachment of the fungus and growth along the lines of epidermal-cell junctions took place, but infection-cushion formation and penetration failed to follow. From all this evidence, Kerr & Flentje have justifiably concluded that both a host type of epidermal surface and a host type of exudate are equally essential for the normal sequence of fungal invasion and infection.

Evidence for the above conclusions has been amplified by Flentje (1957), who showed that invasion of the host took place by

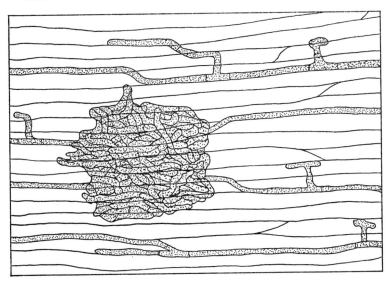

2. Formation of infection cushion by crucifer isolate of *Rhizoctonia solani* on radish hypocotyl. (After N. T. Flentje, 1957.)

means of fine infection pegs (attenuated hyphal tips) from the underside of the infection cushion, a finding confirmed by Christou (1962). Aggregation of hyphae into infection cushions results in a local concentration of fungal inoculum potential above the site of attempted invasion; an infection cushion with its infection pegs can be considered as a single fungal organ, strictly analogous to mycelial strands and rhizomorphs, which again provide a mechanism for the local concentration of inoculum potential (Ch. 4). A similar function for the infection cushions of *Helicobasidium purpureum* (stat. mycel. = *Rhizoctonia crocorum*)

39

has been proposed by Hering (1962), who has observed that penetration of the deeper cortical tissues of the host is effected by a weft of mycelium growing out from the underside of the cushion. The infection cushions produced by *R. solani* are shown in Fig. 2, and those of *H. purpureum* in Fig. 3.

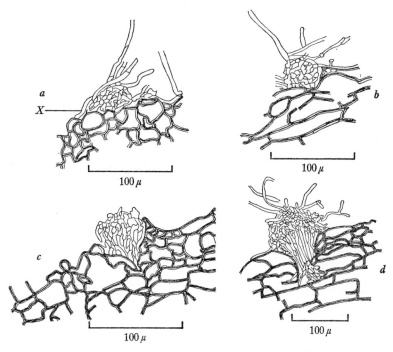

3. Stages in formation of infection cushions by *Helicobasidium purpureum* on carrot: (*a*) young cushion which has broken out of a group of small periderm cells; note two ruptured cell walls (*X*) on left; (*b*) a similar stage, with distension of the underlying cell wall; (*c*) cushion with an elongating peg; (*d*) complete cushion with hyphae radiating from the peg apex. (After T. F. Hering, 1962.)

Flentje has listed four successive stages in which infection failure can occur in host–parasite combinations that do not result in a successful, progressive infection by *R. solani*: (*a*) hyphae fail to become attached to the host surface; (*b*) hyphae become attached to and grow over the host surface, but fail to form infection cushions; (*c*) infection cushions are formed, but penetration by infection pegs is frustrated by thickening of host cell walls as an active response to attempted infection; (*d*) penetration succeeds but hyphae are killed by a hypersensitive reaction of invaded host cells.

Continuing this line of investigation, Flentje, Dodman & Kerr (1963) discovered discrete susceptible areas on the hypocotyl of radish seedlings; these became fewer and eventually disappeared altogether as the seedlings aged. Rubbing hypocotyls with an

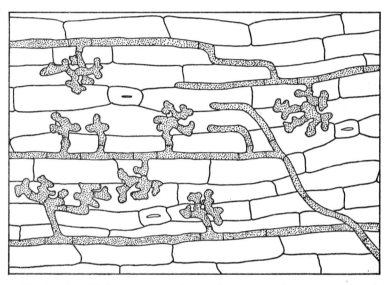

4. Production of lobate appressoria by solanaceous isolate of *Rhizoctonia solani* on tomato hypocotyl. (After N. T. Flentje, 1957.)

organic solvent to remove the wax coating increased the number of infection cushions formed by *R. solani* and returned older, resistant seedlings to the susceptible condition. Earlier workers had often suggested that a wax coating over the plant surface has a protective function, in so far as it causes spore-containing water drops to run off instead of spreading over the epidermis; it now appears that its restriction of host exudation may protect the plant from infection in another way as well.

These findings have recently been extended by Dodman, Barker & Walker (1968 *a*) who, in a study of fifty-three isolates of *R. solani*, found that some of them produced lobed appressoria, as originally described by Flentje (1957), instead of the dome-shaped infection cushions described above (Fig. 4). Isolates from stems and roots usually produced infection cushions, whereas those from foliage usually penetrated either from lobate appressoria through the cuticle or by hyphal growth through stomata. The difference in fungal volume suggests that the inoculum potential generated by an infection cushion is greater than that generated by a lobate appressorium; this could be explained by supposing that resistance to infection offered by stems and roots is generally higher than that offered by leaves, but this speculation needs to be tested against further evidence. In a second paper, Dodman *et al.* (1968*b*) have figured and described these lobate appressoria; they develop as short, swollen side-branches from the leaf-surface hyphae which branch again, often dichotomously. From the underside of each of the hyphal lobes, an infection peg penetrates the cuticle and then enlarges between this and the outer cell wall of the epidermis. They also draw attention to earlier work by Linskens & Haage (1963) demonstrating production of a cutinase by *R. solani*, and suggest that this may assist the process of penetration by fungal turgor pressure. They further suggest that the hypocotyl of radish seedlings is mechanically weaker than that of bean (*Phaseolus vulgaris*) seedlings, because on radish penetration is made by *unconstricted* hyphal tips from the central region of the cushion base, whereas on bean there is mass penetration by the attenuated infection pegs. It seems likely that all these differences in mode of host penetration can be related to the degree of host resistance offered at the various infection sites, but this conclusion must await further evidence.

Lastly, Kerr & Flentje (1957) have made an interesting comparison between the invasion of roots by soil-borne fungal pathogens and infection of leaves by spores of air-borne fungi. In reviewing the latter subject, W. Brown (1936) concluded that there was no evidence for the supposition that a chemical stimulus played any part in the sequence of spore germination and leaf infection, whereas there was ample evidence that a contact stimulus from the surface of deposition would induce attempted penetration; hyphae would readily penetrate artificial membranes of many kinds. Kerr & Flentje point out that once a spore has alighted on a leaf surface, it must infect or perish; unless it is blown, splashed or washed off that surface onto another, there are no other alternatives and most dispersal spores cannot survive for very long. The propagule of a soil-borne fungus, on the other hand, can usually survive in soil, except under extremely adverse conditions, for so long as its reserve of endogenous nutrients suffices to maintain the low respiration rate associated with dormancy. Substrates, in the form of growing roots, travel to root-infecting fungi more commonly than fungi travel or are carried to substrates. For root-infecting fungi, therefore, dormant propagules that germinate only in response to a nutrient stimulus from a potential substrate, such as a living root or a virgin corpus of dead organic material, are behaving in a way possessed of obvious survival value. This was first pointed out by Dobbs & Hinson (1953) after their discovery of the 'general soil fungistasis' that exogenously inhibits germination of fungal propagules in the absence of a nutrient stimulus coming from a potential substrate. Much subsequent work has confirmed the validity of this important generalization (Ch. 7).

Growth of fungal mycelia through the soil

In discussing epidemics of damping-off amongst seedling populations earlier in this chapter, I stated or implied that the mycelia of *Rhizoctonia solani* and of *Pythium* species grew through and over the soil from one seedling to another. This is an established fact and anyone can verify it by direct observation of soil through a stereoscopic microscope. But only a minority amongst soil fungi makes contact with a fresh substrate in this way; most of the species so far investigated lie in wait, as dormant propagules, for substrates to come to them. Substrates for root-infecting fungi are of diverse kinds, and include root exudates as well as root tissues and dead plant material, though not all these fungi are able competitively to colonize all three substrates (Ch. 5). Some years ago (Garrett, 1956a, p. 34) I attempted an inclusive definition of substrates for soil fungi in general: *substrates for soil fungi consist*

43

of living or dead, virgin or partially decomposed plant or animal tissues lying in or upon the soil, and of soluble products diffusing therefrom. The distribution of substrates in soil is discontinuous, inasmuch as the spaces between substrates are filled up with soil mineral particles.

I can summarize my earlier discussion of this subject (Garrett, 1951) by saying that soil fungi secure their substrates in one of two possible ways. Some fungi, exemplified by most species of *Penicillium*, *Aspergillus* and *Fusarium*, direct their nutrient reserves, as exhaustion of the substrate approaches, to the production of spores and other resting propagules. Such fungi do not grow out into the soil as mycelium from the colonized substrate; instead they passively wait for a substrate to come to them. Roots grow through the soil, animals migrate through it, and shoot systems of plants fall upon it; all of these eventually provide substrates for the waiting soil fungi. Other fungi, exemplified by species of *Pythium* and *Rhizopus*, by *Rhizoctonia solani* and by many species of higher basidiomycetes, grow actively through the soil as an expanding, subspherical colony of mycelium from a colonized substrate. There is now little doubt that these two types of behaviour amongst soil fungi are determined by a fundamental physiological difference that is species-specific and was first brought to general attention by Schütte (1956), who separated the fungi that he investigated into two types. *Translocating fungi* are able to translocate nutrients from a food base through an established mycelium traversing a non-nutrient medium; *non-translocating fungi* are unable to do this. In the collection of fungi that he studied, Schütte correlated ability to translocate through an established mycelium with occurrence of protoplasmic streaming in the hyphae. Although subsequent workers have been able to improve upon Schütte's experimental techniques by the use of radio-active tracers (see next section), his general conclusions as to this distinction between translocating and non-translocating species of fungi have been confirmed.

Seedling pathogen as orchid host

A somewhat surprising situation was discovered by Downie (1957, 1959 a, b), when she showed that *Rhizoctonia solani* can act as a suitable mycorrhizal associate for seedlings of *Dactylorchis purpurella* and of several other British orchids. In addition to being a common seedling pathogen, this fungus is a regular saprophytic inhabitant of cultivated soils; this was first demonstrated by Blair (1943), who showed that *R. solani* could continue to grow indefinitely through natural soil containing neither living roots nor fresh organic substrates. Blair therefore suspected that the chief substrate for this fungus might be the cellulose residues in partially decomposed plant material but he was unable to demonstrate that *R. solani* could decompose cellulose, probably because the inoculum potential of his inocula was too low for initiation of cellulolysis. Following upon a report by Tribe (1960) that *R. solani* was a dominant colonizer of cellophane film buried in several cultivated soils, I was able to demonstrate that *R. solani* could decompose filter paper, which is more closely comparable to native cellulose than is the much modified cellophane, in pure culture (Garrett, 1962). I further demonstrated the ability of *R. solani* to colonize filter paper as a competitive saprophyte in unsterilized, natural soil.

Orchid seeds are exceedingly minute and contain insufficient nutrient reserves for establishment and maintenance of the seedling up to the time that it becomes photosynthetically self-supporting. Long before complete proof had been forthcoming, Bernard (1909) and Burgeff (1936) had realized that these seedlings must be *parasitic* upon their fungal partners; this evidence was used by Salisbury (1942) in support of his profoundly important generalization that seed size in flowering plants is determined by the duration of the period for which a young seedling must be maintained by the nutrient reserves of the seed before it becomes photosynthetically self-supporting. Orchid seedlings are nourished by their fungal partners in a reversal of the usual nutritional traffic between fungi and flowering plants; as explained by Salisbury, this permits orchids to produce, out of a limited reproductive capital, very much larger numbers of very much smaller seeds than would otherwise be the case, and thereby to reap the advantage of a wider and more abundant distribution of the species.

This unusual host–parasite relationship has been investigated with modern techniques by S. E. Smith (1966, 1967) working with an isolate of *R. solani* from *Dactylorchis purpurella* and with isolates of other *Rhizoctonia* species from the same host; she showed that these were able both to decompose filter-paper cellulose in pure culture and to colonize it as competitive sapro-phytes in natural soil. Employing radio-active isotopes of phos-phorus and carbon as tracers, she demonstrated translocation of these nutrients through an established mycelium to the infected orchid seedlings. These seedlings were found to maintain a normal growth rate when the only source of carbon was provided by the sugars translocated to them in the mycelium of *R. repens* from a culture on filter-paper cellulose.

For many years the host range of individual isolates of *R. solani* has been known to be limited, and such isolates have been dignified by the name of 'strain', e.g. 'crucifer strain' etc. To any strain of this fungus, therefore, seedlings of some species are resistant and reject the attempted infection; only orchid seedlings, so far as we know, accept infection when suitably offered as an essential for very survival.

DISEASES OF OLDER PLANTS

Diseases of older plants caused by these unspecialized parasites form a somewhat heterogeneous group and many of them are as yet imperfectly understood. In some of these diseases, the pathogen damages the plant simply by a progressive killing of young rootlets as they are produced. An example of this type of disease in an annual crop is the 'no growth' disease of wheat, oats and barley first reported by Samuel & Garrett (1932) to be caused by certain strains of *Rhizoctonia solani* occurring in South Austra-lian soils. In Wisconsin-type soil temperature tanks, we found this disease to be most severe at 12° C and to be progressively mitigated with rise in soil temperature until at 27° C root infection was slight and the check to plant growth was negligible; this explains the natural recovery of winter cereal crops as the soil warms up in spring and early summer. This disease was later described by Hynes (1937) from New South Wales and by Dillon Weston & Garrett (1943) from the county of Norfolk in England. In plantation crops (i.e. bushes and trees), the effects of infection are manifested much more gradually, because these usually older plants have much more extensive root systems, and so such diseases have frequently been given the name of 'decline' or 'slow decline'. Such a sequence of events in the gummosis disease of citrus trees, caused by *Phytophthora citrophthora* (R. & E. Smith) Leonian in Australia, was demonstrated by Fraser (1942, 1949); she showed that the collar canker, beyond which earlier workers had scarcely looked, was the culmination of a fungal attack that started upon the rootlets, working gradually upwards until the resistance of the tree became finally exhausted. In such diseases, infection is limited at least in the early and mid-stages of the disease to immature tissues of the apical region of rootlets, and survival of rootlets therefore depends on disease escape due to local absence of the pathogen. Conversely, the occurrence of progressive rootlet killing on a mature root system can be explained by a locally high population of the fungal pathogen in conjunction with soil conditions that favour infection and so promote a high inoculum potential; if these same soil conditions are also unfavourable for rapid growth and maturation of the rootlets, then conditions optimum for disease development are satisfied in full. For fungal pathogens of this type that happen also to be vigorous competitive saprophytes, incorporation of fresh organic material with the soil, as by ploughing in a cover crop (green manuring), will increase inoculum potential of the pathogen and so aggravate incidence and severity of the disease.

A different explanation has usually to be sought for those diseases in which an unspecialized pathogen can invade mature parts of the root system even during the early stages of disease development. It is indeed possible that a very high inoculum potential of the pathogen will sometimes enable it to infect older parts of the root system that are resistant to a lower and more normal degree of inoculum potential; such effects have been demonstrated in artificially inoculated soils, but their relevance to natural situations is always very questionable. The real primary cause of a disease of this type is usually some *predisposing factor* of the soil or subaerial environment that prevents older regions of the root system from developing normal mature-tissue resistance to infection. Both nutrient deficiencies and soil toxins may act on the plant in this way; for development of the disease, the predisposing factor is quite as essential as the fungal pathogen.

Factors favouring the pathogen: fresh plant material buried in the soil

If green manure crops, weeds or other fresh plant material are ploughed into the soil, there is always some danger that such material will provide a suitable substrate for these unspecialized pathogens. Green and still living plant tissues will selectively favour *infection* by pathogens, because residual host resistance will at first retard *colonization* by obligate saprophytes; in this situation, the difference between parasites that infect still living tissues and obligate saprophytes that can colonize only dead tissues becomes more than a merely academic distinction of terminology. In such a situation of disease risk, dead crop residues will be a safer soil supplement than a green manure crop for maintenance of soil organic content. *Pythium* species in particular are common colonizers of fresh green plant material; thus Sawada, Nitta & Igarashi (1965) found that incorporation of fresh, green clover leaves with the soil resulted in a multiplication of *Pythium* species, which then infected oat seedlings sown immediately after the green manuring. Trujillo & Hine (1965) observed that fresh papaya residues buried in soil were colonized by *Pythium aphanidermatum* and *Phytophthora parasitica* within 48 hours, leading to a seedling root-rot in papaya nurseries. In soils having a negligible population of these pathogens, there was no harmful effect of the green manuring, showing that in this case phytotoxic products of decomposition were not a predisposing factor in the aetiology of the seedling disease.

Diseases of older parts of the root system and of the collar region sometimes result from massive colonization by fungal pathogens of dying or dead plant material in close contact with the crop plants. Tropical species of *Rosellinia* can cause a particularly troublesome disease of plantation crops such as cacao in situations where high rainfall and dense shade thrown by the crop encourage accumulation of leaf litter. As Waterston (1941) has described, these fungi grow over and through the surface litter as an advancing sheet of mycelium, and from this food base these fungi invade trunks as well as roots of the plantation crop; the strongest trees are as susceptible to the disease as the weakest ones.

Southern blight of peanuts (*Arachis hypogea*) in Virginia and the southern U.S.A. caused by *Sclerotium rolfsii* is similarly favoured by buried plant residues, on which the pathogen multiplies (Boyle, 1961; Garren, 1961). Residues of the previous crop, or of a cover crop, and living weeds buried under the soil surface all act as food bases for *S. rolfsii*. Garren (1964) has obtained greatly improved control of southern blight by two changes in crop husbandry: (1) deep ploughing substituted for shallow ploughing, so that plant residues are buried as far as possible from the upper root system, around which *S. rolfsii* is most active; (2) control of weeds by herbicides substituted for control by burying beneath the soil surface.

Factors predisposing the host to infection

Nutrient deficiencies. Any nutrient deficiency, if sufficiently severe, will seriously impair host resistance to infection; correction of the deficiency will control the disease and so infection by one or more fungal pathogens is the result of an unhealthy condition rather than the primary cause of it. On the Canadian prairies, most serious losses were at one time caused by browning root-rot of cereals, with which was associated infection by various species of *Pythium*. Vanterpool (1940) has summarized his own researches, which led to the discovery that phosphate deficiency was the primary cause of the diseased condition. Since then, Vanterpool (1952) has discussed reasons for the virtual disappearance of this one-time perplexing disease; the now adequate use of phosphatic fertilizers has been supplemented by crop husbandry measures for control of soil erosion and conservation of soil phosphorus.

I can claim some understanding of the work carried out by the pioneers of root-disease investigation, not only through their papers but also because I have myself repeated most of their mistakes, though with less excuse for such blunders. Nearly 40 years ago, I spent some time in fruitlessly investigating the poor growth of young cereal crops on an area in South Australia; I isolated some unspecialized fungal pathogens from infected roots, but was unable to reproduce the disease in inoculation trials, for which I innocently employed sterilized soil of another type. The answer to this particular problem was found by Millikan (1938). working on a similar soil type in the adjoining State of Victoria; it was a dressing of zinc sulphate to correct a growth-limiting deficiency of this minor nutrient. I was pleased at least to find, however, that Millikan had recorded much the same range of unspecialized fungal pathogens as I also had isolated from infected roots.

Phytotoxins in decomposing plant residues. Decomposing plant residues have many and diverse effects upon soil and the plants growing therein, but in general the beneficial greatly outweigh the harmful effects of these decompositions on crop growth. Mineral nutrients, of which the major one is nitrogen, are released and converted by the soil microflora into forms suitable for uptake by plant roots; the residue of decomposition, i.e. humus, improves the tilth and water-holding capacity of the soil. Control of some soil-borne diseases can be obtained by incorporation with the soil of a sufficient volume of crop residues of the right type (Ch. 8); a variety of effects upon individual diseases is known to follow organic manuring, though not all these beneficial effects strictly merit the term of 'biological control' (Garrett, 1965).

Nevertheless, the effects that follow green or other organic manuring are not invariably beneficial, and much depends upon the composition of the crop residues. Incorporation with the soil of organic matter having a high C/N ratio, such as mature cereal straw, results in a complete locking-up of soluble soil nitrogen within diffusion reach of the straw in the cells of the decomposing micro-organisms. Until decomposition is completed and the microbial nitrogen is mineralized after exhaustion of the carbon substrate, the decomposing straw is depleting the soil of available nitrogen, so that the first crop after straw manuring will suffer severely unless the deficiency is corrected by an adequate dressing of nitrogenous fertilizers; such a shortage of nitrogen will aggravate the severity of various diseases, including take-all of cereals (Garrett & Mann, 1948).

Once anticipated, a difficulty like temporary nitrogen deficiency of the soil, consequent upon straw manuring, can easily be over-come; a more difficult problem is presented by the formation of substances phytotoxic to crop plants during decomposition of plant residues. Some years ago, I reviewed earlier work on the Ontario root-rot of strawberry plants, a disease of complex and uncertain aetiology (Garrett, 1956a, p. 120). None of the fungal pathogens that had been isolated from the root systems of diseased plants were sufficiently specialized parasites to be likely to cause, unaided, a continuing and severe disease of mature plants; the most pathogenic were species of *Pythium* together with *Rhizoctonia solani*. Cover crops of maize, red clover and timothy grass made the disease worse; cover crops of soybean improved the condition of the plants on strawberry-sick soils, as also did acid-producing fertilizers like glucose, acetic acid and dried blood. At that time I suggested, with reference to the then recent work of Cochrane (1948, 1949), that the primary cause of this root-rot might be the toxicity of certain decomposing crop residues towards the straw-berry plant, and that infection of the roots by a variety of unspecialized fungal pathogens was a consequence rather than a primary cause of their unhealthy condition. Work published since that time on the temporary phytotoxicity of various decomposing crop residues has increased the probability that Ontario root-rot of strawberries is a toxin-induced disease.

The whole subject of phytotoxins in relation to soil-borne diseases has been recently reviewed by Börner (1960) and by Patrick & Toussoun (1965). A sample of the recent evidence for the production of phytotoxins is as follows. Patrick, Toussoun &

Snyder (1963) studied this problem both in the laboratory and glasshouse and in the field in the Salinas Valley in California, where injuries to seedlings of lettuce, bean (*Phaseolus vulgaris*), spinach, broccoli and tobacco followed cover crops of wheat, barley, rye, Sudan grass, vetch, bean (*Vicia faba*) and broccoli. Injuries to seedlings included poor germination, retarded growth of young tap roots, and discoloration, necrosis and death of apical meristems. The effect was local and proximal contact with decomposing plant fragments was required for injury to seedling tissues. Fungi isolated from lesions were mostly non-pathogenic. Water extracts of decomposing plant tissues showed maximum toxicity within the period 10–25 days from start of decomposition; non-toxic or even growth-stimulating extracts were obtained after 30 days. In Canada, Patrick & Koch (1963) observed that exposure of tobacco roots to water extracts of decomposing rye and timothy residues increased their susceptibility to black root-rot caused by *Thielaviopsis basicola* (Berk. & Br.) Ferraris, and even caused a breakdown in the resistance of tobacco varieties normally resistant to this disease. Increased susceptibility to foot-rot of beans, caused by *Fusarium solani* (Mart.) Sacc. f. *phaseoli* (Burkh.) Snyder & Hansen, was similarly promoted by water extracts of various decomposing crop residues (Toussoun & Patrick, 1963). Drops of the phytotoxic extracts placed on bean hypocotyls led to increased exudation of amino acids and other nitrogenous compounds, which had earlier been shown by Toussoun, Nash & Snyder (1960) to promote lesion production by this pathogen. Root-rots of bean caused by *Thielaviopsis basicola* and by *Rhizoctonia solani*, respectively, were also much exacerbated by exposure of root systems to extracts of decomposing crop residues.

Quite recently, progress in identifying some of the phytotoxic chemicals has been reported by Toussoun, Weinhold, Linderman & Patrick (1968). They employed as raw material the green tissues of barley, cotton, cowpea and soybean, which gave phytotoxic water extracts 7–10 days after burial in soil. Toxicity of the extracts reached a maximum after 3 weeks and declined after 6–7 weeks from first burial. Substances soluble both in water and in ether accounted for some 60% of the total toxicity. The major toxic components were benzoic and phenylacetic acids; also identified were hydrocinnamic (3-phenylpropionic) and 4-phenylbutyric acids. Linderman & Toussoun (1968*a*) followed this up by testing for a possible predisposing effect of treatment with hydrocinnamic acid on 3-day-old cotton seedlings; the tap roots were immersed in a 75 ppm solution for 2 h, and the seedlings were then replanted in soil infested with chlamydospores of *Thielaviopsis basicola*. Pretreatment with hydrocinnamic acid increased the disease rating of the tap roots from 66% in the control series (water alone) to 96% in a disease-susceptible variety of cotton; in two disease-resistant varieties, the mean disease rating for the two was increased from 5 to 32% by the treatment. Linderman & Toussoun (1968*b*) analysed the disease-promoting effect of pretreatment with these toxins into two components: (1) many more chlamydospores of *T. basicola* germinated at the root surface after pretreatment and gave rise to lesions; (2) individual lesions expanded further on roots receiving the pretreatment than on those of control plants. Neither the phytotoxic ether extract of decomposing barley tissues, nor any of the four component organic acids, were found to stimulate chlamydospore germination in soil apart from roots; the authors have therefore suggested that increased percentage germination and infectivity of the chlamydospores may follow an increase in volume and/or nutrient content of root exudate induced by the pretreatment.

It now seems probable that phytotoxins of similar origin play an important part, perhaps the leading part, in the so-called 'replant diseases' of apple, peach and citrus, where it is often difficult to establish a new plantation on the site of an old one of the same species. Börner (1959) has suggested that toxic decomposition products of phlorizin, a natural constituent of apple-root bark, may be responsible, at least in part, for the apple replant disease. The peach replant problem has been studied with great thoroughness by Z. A. Patrick and his associates at Harrow, Ontario, and all this work has been summarized by Patrick & Toussoun (1965). Peach root bark and other tissues were shown to contain relatively large amounts of the cyanogenic glucoside, amygdalin, along with the appropriate hydrolysing enzyme, emulsin. Amygdalin is not itself toxic to peach roots, but soil micro-organisms can decompose it with liberation of two highly toxic degradation products, benzaldehyde and hydrogen cyanide. So can the enzyme systems of the nematode *Pratylenchus penetrans* (Cobb) Filip. & Stek., which is present in large numbers in the peach soils of south-western Ontario. The mechanical damage to living root tissues following nematode penetration is of itself sufficient to release the toxic degradation products, by bringing together in disorderly fashion the amygdalin and the emulsin already present, though apart, in healthy peach tissues. Thus the work of Patrick and his associates seems to have unravelled a complex chain of cause and effect in this peach replant disease.

Diseases of this type present the most difficult problems now to be found in the whole field of root-disease investigation. The pioneering days are thus not yet over, but the position of a pioneer is perhaps more enjoyable in retrospect than in being.

3   Specialized parasites: vascular wilt fungi

The vascular wilt fungi constitute an interesting and economically important group of specialized, soil-borne pathogens. Taxonomically, they are comprised within formae speciales of *Fusarium oxysporum* (Snyder & Hansen, 1940) and the genera *Verticillium* and *Cephalosporium*. Amongst species of *Verticillium*, the most important as wilt pathogens are *Verticillium albo-atrum* Reinke & Berth. and *Verticillium dahliae* Kleb.; plant pathologists in the U.S.A. usually include *V. dahliae* within *V. albo-atrum*, although there are differences both morphological and in behaviour between the two pathogens, and this difference in nomenclatural practice can lead to confusion. Diseases caused by these pathogens are widespread and exceptionally difficult to control except by the use of disease-resistant species or cultivars.

In an earlier account of these fungi (Garrett, 1956a, p. 52), I suggested that they had more affinities with the unspecialized than with the specialized parasites, though on existing evidence they did not seem to conform well with the criteria for either group. My earlier view was based partly on what I now think to be an erroneous interpretation of the host–parasite relationship of vascular wilt fungi, but more particularly on the long survival of these pathogens in soil apart from their host plants (up to 10 years or more). Saprophytic survival in dead infected tissues of host plants could not account for so great a longevity, nor was it generally appreciated at that time that chlamydospores (both hyphal and conidial) produced by formae speciales of *Fusarium oxysporum* constitute a highly efficient mode of dormant survival in soil (Ch. 7). For these reasons, I concluded that these fungi were likely to survive in soil apart from their host plants as free-living competitive saprophytes, a conclusion compatible with the view that they were comparatively unspecialized parasites. Neither of these conclusions is now tenable for formae speciales of *Fusarium oxysporum*; their relatively narrow host range as pathogens and more recent knowledge on their host–parasite relationship, to be discussed later, both suggest that they can now be regarded as specialized parasites. This conclusion accords well with the fact that satisfactory evidence is now seen to be lacking for the supposition that they can survive indefinitely in soil as free-living saprophytes, as at least some unspecialized parasites can do. As a result of experiments with *F. oxysporum* f. *cubense*, Stover (1962) concluded that this fungus was not an effective competitive saprophyte in the soil; since then, he has decided that the great longevity of this pathogen between successive banana plantations is due to its production of chlamydospores in the infected tissues (Stover, 1969).

THE VASCULAR INFECTION HABIT

In a susceptible species or cultivar of crop plant, the infection sequence appears generally to proceed as follows. Actual or proximal contact with a young root or rootlet (small branch root) stimulates, through diffusion of nutrients in root exudate, germination of the fungal inoculum. This inoculum may consist of resting propagules of the pathogen, such as chlamydospores of fusaria and microsclerotia of *Verticillium dahliae*, of a dying or dead infected root, or of dispersal spores produced at the surface of the latter and perhaps carried away from it by movement of soil water (Sewell, 1959). Infection of young roots and rootlets occurs through the immature cortex of the apical region and is thus similar, and similarly limited, to infection by unspecialized fungal pathogens (Ch. 2). Mature parenchyma cells, whether of cortex, pith or xylem tissues, seem to be resistant to infection; in this respect, vascular wilt fungi show a resemblance to unspecialized pathogens and this resemblance contributed to my earlier view of their host–parasite relationship, now recanted. Having penetrated through the cortex of the root apical region, the fungus then passes through the endodermis, which in this region is not yet fully differentiated and thus offers incomplete resistance to passage of the fungus into the protoxylem vessels. From these, the fungus migrates upwards into the older xylem tracts, and often proceeds to the top of the aerial shoots and even into the petioles and main veins of the leaves. Once such a vascular wilt parasite has established a progressive infection of the xylem tracts, any further resemblance between it and an unspecialized pathogen disappears. This is well illustrated by the recent work of Beckman (1966), who found that even if a mixture of soil organisms from root washings was introduced into the vascular cylinder of tomato tap roots, the infection was quickly suppressed by host resistance and did not get far. This and other evidence demonstrate an important distinction between specialized vascular wilt parasites and the class of unspecialized pathogens; the mechanism of this active resistance will be discussed later in this chapter.

It is now firmly established that movement of some of these vascular wilt fungi from the root system to the top of the shoot system is much more rapid than can be explained solely by upwards mycelial growth through the xylem vessels, as was formerly supposed. This rapid upwards movement is effected by carriage of microconidia in the transpiration stream, and has been demonstrated by Trujillo (1963) for *Fusarium oxysporum* f. *cubense* in the banana plant. He observed that the appearance of the fungus in the xylem vessels was yeast-like, with abundant microconidia but sparse mycelium. The microconidia appeared to bud profusely in the xylem sap, producing long, branched chains of further microconidia. These microconidia were transported upwards through each xylem vessel by the transpiration stream; arriving at the end wall of a vessel, the microconidia germinated and the germ tubes penetrated the pit membranes and thereafter budded off more microconidia on the further side of the end wall opening onto the lumen of the next vessel in the file. In this way, *F. oxysporum* f. *cubense* was found to migrate from the rhizome to the top of the pseudostem of a 25 ft tall banana plant in less than 2 weeks.

Similar evidence has been produced by Sewell & Wilson (1964) from a study of the movement of *Verticillium albo-atrum* in the hop plant. The concentration of the small conidia of this fungus in the xylem sap was found sometimes to reach a level of nearly 7×10^3/ml. Many field botanists may have reflected upon the significance of the exceptionally high growth rate of annual stems of many climbing plants, which is a characteristic of paramount value to their mode of life; it is thus of interest to note that Sewell & Wilson found continuous open vessels in the hop bine reaching a length of 150 cm in 4-week-old stems, and they have further observed that by early summer such continuous open vessels extended almost to the top of field-grown bines. In such open vessels, no end walls obstruct the upwards travel of the conidia, which were found to reach 20 cm above the injection point in 24 hours, up to 150 cm in 7 days, and up to the top of the bine within 14 days. Similar data have been obtained by Presley, Carns, Taylor & Schnathorst (1966), who inoculated stems of glasshouse-grown cotton plants averaging 115 cm in height with a conidial suspension (10^6/ml) of *V. albo-atrum*. The fungus was isolated from the topmost petioles of the inoculated plants after only 24 hours.

THE VASCULAR WILT SYNDROME

A variety of macroscopic symptoms goes to the make-up of the typical vascular wilt syndrome. Early and mid-stages of the disease are usually characterized by intermittent wilting, due to loss of turgor as in drought wilting; a later and more typical symptom is desiccation of leaves in acropetal succession towards the stem apex, usually preceded by leaf yellowing. Other symptoms are epinasty (downwards curvature) of petioles, production of adventitious roots from the lower part of the stem, and a thickening of the stem due to prolonged or renewed cambial activity in the production of secondary xylem.

This wide variety of symptoms suggests that several mechanisms are concerned in their production as a consequence of vascular infection. The mechanisms of the vascular wilt syndrome have attracted much research over the last decade, and for a detailed discussion of all this work I can recommend Chs 9 and 10 in Wood's (1967) recent book. He has classified possible causes of wilt symptoms, including those produced by bacterial vascular pathogens, as follows:

(1) Production by the pathogen of low molecular weight toxins, amongst which lycomarasmin and fusaric acid have received most attention. It is now conceded that lycomarasmin does not fulfil the conditions for its acceptance as a *vivotoxin*, i.e. a chemical agent that is produced in sufficient quantity within the infected tissues to cause the observed symptoms (Dimond & Waggoner, 1953). Evidence for the implication of fusaric acid as a vivotoxin, however, is considerably more convincing. Wilting can be caused either by a reduction in uptake and transport of water, e.g. by mechanical obstruction of the xylem tracts, or by increased loss of water due to an increase in the permeability of cell membranes or to an impairment of stomatal functioning. It appears that fusaric acid contributes to wilting by an effect on cell membrane permeability.

(2) Production by the pathogen of metabolites having growth-regulating activity for higher plants, such as ethylene and indole-acetic acid (IAA), which can cause epinasty of petioles and production of adventitious roots.

(3) Production by the pathogen of polysaccharides and other substances of high molecular weight, which can cause impedance of water flow in the xylem.

(4) Vascular obstruction by gels resulting from degradation of xylem cell walls by enzymes of the pathogen, and chiefly by pectinase enzymes.

UNDERGROUND SPREAD OF INFECTION

The first plant pathologists to study thoroughly the distribution of infection in plants suffering from a vascular wilt disease found that infection, and the tissue discoloration resulting from it, was virtually confined to the xylem tracts, at least up to the penultimate phase of the disease. It was further observed that, in a suitably humid atmosphere, a vascular wilt fungus would emerge to the outside of the stem and the leaf petioles and sporulate thereupon when death of the plant from the disease was approaching, though not before. Such observations can be interpreted by supposing that the mature parenchymatous tissues surrounding the vascular tracts are resistant to infection and that this resistance breaks down only with the advent of disease-induced senescence in the whole or part of the infected plant. This confinement of the infection within the xylem tracts until the penultimate phase of disease development carries an important corollary concerning the underground spread of vascular wilt diseases from plant to plant. If the fungus does not break out of the vascular tracts until disease-induced senescence has set in, then an infected root system does not become *infective* to those of neighbouring healthy plants until at least some of its roots are dying from the effects of the infection.

In an experiment to test this assumption that an infected root does not become infective to other roots in contact with it until it becomes moribund, Roberts (1943) found that approximately twice as many tomato plants became infected in a given time around Verticillium-infected plants that had died as a result of bark-ringing as became infected around similar Verticillium-infected plants that had not been ringed and still remained alive. A similar but more extensive experiment was carried out by Isaac (1953) with five species of *Verticillium*, of which two were severe vascular pathogens and three were mild ones. Observations were made in two consecutive seasons on spread of these five wilt diseases from a central, wound-inoculated plant to two concentric circles of uninoculated plants set around it. The two severe pathogens, *V. albo-atrum* and *V. dahliae*, infected 3–4 of the 6 inner-circle plants during the first season, and 2–5 of the 12 outer-circle plants during the second season. The three mild pathogens, *V. nigrescens* Pethybr., *V. tricorpus* Isaac and *V. nubilum* Pethybr., failed to infect any of the surrounding plants during the first season, and infected at most only a single plant of the inner circle during the second season. In his interpretation of these results, Isaac has suggested that both death of the root system and development of its infectivity to those of neighbouring plants are much delayed by the mildness of the vascular infections caused by the second group of three pathogens, as compared with earlier killing by the first group of the two severe pathogens. Isaac's evidence has a very direct bearing on the question at issue, and his interpretation is compatible with what other evidence we have on this matter.

A convincing visual demonstration of this postulated sequence of events has been made by Sewell (1959), employing glass observation boxes for a study of the development of *Verticillium albo-atrum* in and on the root systems of soil-inoculated tomato plants; he was able to observe and photograph sporulation of this fungus on roots *in situ* at the soil–glass interface. Sporulation did not begin until degeneration of the root system had become apparent to the naked eye; both root degeneration and fungal sporulation began first amongst the fine roots, and these events followed in turn with the main lateral roots and the tap root. In an experiment to determine the effect of root-system killing by various treatments on time of onset of sporulation, Sewell obtained a close correlation between rapidity of root-system killing and precosity of sporulation by *V. albo-atrum* (Table 2).

Table 2. *Sporulation of* Verticillium albo-atrum *on infected tomato roots following treatments differentially affecting their death rates*

(From Sewell, 1959)

Time after treatment (days)	Soil drenched with 1 % solution of sodium chlorate	Shoot system removed at ground level	Stems ring-barked 1 in above soil	No treatment
10	—	—	—	—
20	+ + +	—	—	—
30	+ +	—	—	—
40	+ +	+ + +	—	—
54	—	+ +	—	—
60	—	+ +	+	—
70	—	+ +	+ + +	—
86	—	+	+ +	+ + +
94	Not recorded		+	+ +

Density of sporulation: +, low; + +, medium; + + +, high.

In his general discussion of these results, Sewell has pointed out that spread of infection from infected to healthy root systems can be mediated by the mass transfer of conidia from sporulating root surfaces in moving soil water, and is thus not entirely dependent upon actual or proximal contact between infective and healthy roots, as I had formerly supposed (Garrett, 1956*a*, p. 54). In this conclusion he is clearly correct, as many experiments have shown that conidia of *V. albo-atrum* can act as infective propagules.

Amongst diseases caused by root-infecting fungi, the vascular wilts constitute the only group causing a systemic infection of the whole plant, from the rootlets to the petioles and veins of the leaves; it is at least possible that such a general systemic infection can originate from a single infection focus in the soil. This group of diseases is thus not unlikely to obey Van der Plank's (1947) generalization for systemic virus diseases, which states that the larger the catchment area offered by an individual plant for insect vectors of the virus, the greater is the probability of infection, and the higher will be the percentage incidence of disease over a given area. If plants are closely crowded, therefore, the *proportion*, though not the actual number, that becomes infected by insect vectors over a given area is likely to be lower than if they are widely spaced. Such a relationship is likely to hold true only below certain limits of insect vector population, and similarly only below certain limits of wilt pathogen population in the soil. Nevertheless, the comparison is of much interest and suggests that close planting might effect a reduction in loss from vascular wilt diseases in favourable circumstances, just as it has been found to do for some insect-transmitted, systemic virus diseases.

MECHANISMS OF WILT RESISTANCE

Until about 25 years ago, it was generally believed that the quality of resistance or susceptibility characteristic of a particular crop cultivar in its reaction to a particular wilt pathogen was possessed equally by root and shoot systems. This prevailing assumption was then challenged by Heinze & Andrus (1945), who performed grafting experiments with tomatoes of high resistance (Pan America) and low resistance (Bonny Best) to *Fusarium oxysporum* Fr. f. *lycopersici* (Sacc.) Snyder & Hansen; all four possible stock-scion graft combinations were constituted for inoculation through the roots. The results were clear-cut; both Bonny Best and Pan America stems were susceptible to infection on Bonny Best (low resistance) root systems, and resistant on Pan America (high resistance) root systems. Heinze & Andrus therefore concluded that the resistance of Pan America was confined to the root system, and that the stems of Pan America, as well as both stems and roots of Bonny Best, were fully susceptible to infection.

This finding was followed up by Keyworth (1953), who performed a similar type of grafting experiment with varieties of hop showing high and low resistance, respectively, to a virulent race of *Verticillium albo-atrum*, with which they were inoculated via the roots. Results essentially similar to those of Heinze & Andrus with *Fusarium oxysporum* f. *lycopersici* were obtained; severe wilt developed in both types of scion stems when grafted on to roots of low resistance, whereas only mild wilt developed in either type of scion stem when growing on roots of high resistance. Keyworth then tried the effect of direct stem inoculation by two methods; for this purpose he employed the available alternative host–parasite combinations to give mild and severe forms of wilt, viz. mild and virulent races of *Verticillium albo-atrum* on the low-resistance variety Fuggle. By the first inoculation method, a drop of spore suspension was injected with a hypodermic syringe directly into the xylem tissue of the stem; only a very small amount of woody tissue could be thus injected with the spore suspension, owing to the solidity of the stem. Although infection of the xylem was obtained by this technique, and visible discoloration of the wood often extended for several feet or more above the point of inoculation, yet only typical symptoms of mild wilt developed, whether the inoculant race of *V. albo-atrum* was 'mild' or 'virulent'. By the second inoculation method, a substantial dose of inoculum, consisting of some 1–2×10^5 spores of either race of the fungus, was suspended in $0 \cdot 5 \%$ water agar and injected into an internodal pith cavity. This method of inoculation produced severe wilt symptoms above the point of inoculation, both with 'mild' and with 'virulent' races of the fungus. In these stem inoculation experiments, therefore, the type of disease produced

63

depended upon the type of inoculation, and was not affected by the fungal race employed.

In Keyworth's second experiment, the small dose of inoculated spores may be taken to represent the low infection potential of the pathogen coming up to the stem base from an infected root system of high resistance; the large dose of inoculated spores represents the high infection potential of the pathogen coming up from an infected root system of low resistance. This second experiment by Keyworth enabled him to make an important advance in understanding the mechanism of resistance. Unlike Heinze & Andrus, who had concluded that stems, whether of low- or high-resistance plants, had only a low resistance to infection, Keyworth was able to demonstrate that stems even of a low-resistance variety possessed a significant degree of resistance to infection. To be more precise, Keyworth demonstrated that the degree of resistance possessed by stems of a low-resistance variety was adequate to check infection from a high-resistance root system, but not that coming from a low-resistance root system. Talboys (1958a, b, 1964), who continued Keyworth's line of investigation at the East Malling Research Station, has developed a general hypothesis that can account for varying degrees of resistance to vascular wilts without the need to invoke any resistance-differential in the stems. Talboys has postulated two phases in the development of a vascular wilt disease. The *determinative phase* comprises the whole sum of root cortex infections that get through into the xylem; this is determined first by the inoculum potential of the pathogen in the soil, and secondly by the degree of resistance to infection possessed by the extravascular tissues, i.e. cortex, endodermis and pericycle, of the particular host species or cultivar. If the sum of primary infections of the xylem is sufficient to overcome the resistance of the xylem tissues, i.e. vessels with associated parenchyma cells, a severe manifestation of the wilt disease follows; if it is not sufficient, then only mild symptoms of wilt can develop, or the infection is suppressed without manifestation of any symptoms at all. Symptom expression, or the lack of it, is distinguished by Talboys as the *expressive phase* of the wilt disease. Talboys's hypothesis thus postulates that the early stages of infection of the root system, up to and including first colonization of the xylem, are critical for the subsequent development of a wilt disease; the total volume of mycelium colonizing the xylem has to reach a critical value for xylem infection to become progressive, otherwise it will be halted by the xylem defence-response. This xylem defence-response is postulated as being much the same in different species and cultivars of plants, inasmuch as it is a generalized response to physical damage and casual infections; the differential in host resistance to vascular infections is constituted by a difference in response of the extravascular tissues in the early stages of infection. After this preliminary adumbration of Talboys's hypothesis, I shall now discuss the evidence for resistance differentials and the mechanism of the generalized xylem defence-response.

First, it is possible, though not yet firmly established, that the degree of resistance to vascular wilt infection possessed by a species or cultivar may sometimes manifest itself even in the pre-penetration stage of infection. It is not unlikely that the composition of root exudates sometimes reflects, at least to some extent, the composition of the plant's cell sap, and so we might expect sometimes to find a differential effect of root exudates from different hosts on fungal spore germination. Such a discovery was indeed claimed by Buxton (1957), who reported an effect of this sort with different races of *Fusarium oxysporum* Fr. f. *pisi* (Lindf.) Snyder & Hansen and several pea cultivars varying in their susceptibility to the different races of the pathogen. Unfortunately, however, Kommedahl (1966) subsequently reported complete failure to confirm Buxton's findings. Before this, Buxton (1962) had also reported that root exudates of the Lacatan banana, resistant to Panama disease, reduced percentage spore germination of *Fusarium oxysporum* f. *cubense*, whereas root exudates of the susceptible Gros Michel did not.

Although a significant effect of root exudates on the pre-penetration stage of infection thus needs to be confirmed, there seems no reason to doubt that the progress of cortical infection is checked by resistance mechanisms in young roots of resistant cultivars. Thus Talboys (1958*a*), in a study of 'wilt-sensitive' (= susceptible) and 'wilt-tolerant' (= somewhat resistant) varieties of hop inoculated with mild and virulent strains, respectively, of *Verticillium albo-atrum*, observed that in relatively resistant varieties, infection by the fungus was hampered by cell-wall lignification in the epidermis and cortex, and that penetrating hyphae were sometimes occluded and arrested by the proliferation of lignitubers as outgrowths from the host cell walls; suberization of the endodermis was also more precocious in the resistant than in the susceptible varieties. In a

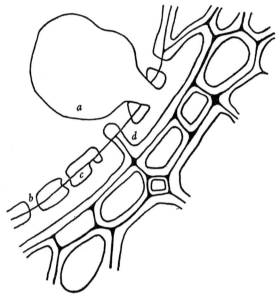

5. Development of a tylose in a vessel of oak infected by *Ceratocystis fagacearum*, (*a*) tylose; (*b*) pit-closing membrane; (*c*) secondary wall; (*d*) ray cell. (After Struckmeyer, Beckman, Kuntz & Riker, 1954.)

further study, on the vascular phase of infection, Talboys (1958*b*) observed that acute symptoms of hop wilt were associated with intense development of mycelium in the xylem vessels but sparse production of tyloses; conversely, mild symptoms were associated with sparse mycelium but intense tylosis of the vessels. Tyloses are balloon-like outgrowths into the lumen of a vessel produced by xylem parenchyma cells of medullary rays that are in contact with the vessel through a perforation or 'pit' in its secondary wall;

the parenchyma cell is separated from the lumen of the vessel only by a thin membrane, and it is by extension and growth of this pit membrane into the vessel lumen that a tylose is produced. This process has been very clearly explained and figured in a paper on the oak wilt disease caused by *Ceratocystis fagacearum* (Bretz) Hunt in the U.S.A. by Struckmeyer, Beckman, Kuntz & Riker (1954); Fig. 5 is reproduced here from their paper.

In hops showing only mild symptoms of wilt, most of the infected vessels are occluded by tyloses; as Talboys has pointed out, however, this loss of conducting capacity in infected plants is compensated by the rapid development of new secondary xylem. This results in a symptom of mild wilt known as 'thick bine'. In monocotyledons, such as the banana when infected by *Fusarium oxysporum* f. *cubense*, there is no mechanism for the production of secondary xylem to offset closure of vessels by tyloses, so we can only suppose that the plant possesses surplus conducting capacity.

In the discussion of his results, Talboys (1958*b*) suggested a simple explanation of the inverse correlation he had found between density of mycelium and frequency of tyloses in infected xylem vessels of the hop plant, by postulating that a low concentration of fungal metabolites in the xylem stimulates tylosis but that a high concentration inhibits it. This explanation fits not only his own observations on infected hops, but also more recent ones by C. H. Beckman and his associates on the Panama disease of bananas. Thus Beckman, Halmos & Mace (1962) observed that temporary vascular occlusion by gel, followed by permanent occlusion by tyloses, was rapid and continuous in the resistant banana Lacatan, at all three growth temperatures studied (21, 27 and 34° C). In the susceptible Gros Michel, by contrast, vascular occlusion was poor at both 21 and 27° C; the gel disappeared rapidly and tylosis was slow. Only at 34° C was vascular occlusion rapid and effective, and this was paralleled by a high degree of wilt-resistance in Gros Michel grown at this higher temperature. Growth and sporulation of *F. oxysporum* f. *cubense* in the vascular sap were found to be of a similar order in both Lacatan and Gros Michel; the resistance of Lacatan at all temperatures and of Gros Michel at 34° C was ascribed to obstruction of upwards spore movement in the vessels occluded by tylosis. This conclusion was confirmed by observations on the movement of spore-sized red vinyl particles injected into the vessels. But although vascular occlusion by Gros Michel was insufficient, except around 34° C, to halt the upwards progress of *F. oxysporum* f. *cubense*, yet it was found adequate to terminate infection by pathogens not specialized to vascular development in this host. Thus Beckman & Halmos (1962) introduced into the xylem of Gros Michel roots two vascular pathogens of other hosts, *Fusarium oxysporum* f. *lycopersici* and f. *pisi*, and two non-vascular pathogens, *Fusarium solani* and *F. roseum*; all four at first grew well enough in the xylem, but were later immobilized by vascular occlusion.

These experiments by Beckman and his associates seem to have settled quite definitely, at least for the Panama disease of bananas, a question arising out of Talboys's general hypothesis for the development of vascular wilt diseases, discussed above. This hypothesis disclaims, as superfluous, any difference in the xylem defence-response between host species and cultivars, and postulates that the differential is constituted chiefly by the varying infection-responses of the extravascular tissues. Now in the experiments by Beckman and his associates just described, inoculation was by means of *a standard dose of microconidia* introduced directly into the xylem vessels. This constitutes convincing evidence that there is a highly significant difference between the xylem defence-responses of Lacatan and Gros Michel bananas. For the banana at least, and perhaps for some other host species as well, Talboys's 'determinative phase' must be extended further up the infection sequence and well into the xylem. To say this is not to deny that early stages in the infection sequence for *F. oxysporum* f. *cubense* are of critical importance for subsequent development of the disease; evidence for this is afforded by the work of Rishbeth (1955), to be discussed in the next section.

Beckman & Zaroogian (1967) have investigated the gels that occlude banana vessels in advance of tylosis, and report that they are primarily composed of pectins, calcium pectates, hemicelluloses and traces of protein; they suggest that these gels result from enzymic degradation and swelling of the perforation plates, end walls and side walls of vessels. In a study of cell-wall composition in two susceptible and two resistant varieties of banana, the only correlation with the character of resistance found by Zaroogian & Beckman (1968) was a significantly higher content of one hemicellulose out of two found in the walls. Whereas it seems highly probable that fungal enzymes are responsible for the production of these gels, the fungal metabolites responsible for induction of tylosis have not yet been identified with any certainty. Mace & Solit (1966) have reported, however, that 3-indoleacetic acid (IAA) induces tylosis of banana vessels, and they refer back to the demonstration by Mace (1965) that *F. oxysporum* f. *cubense* synthesizes IAA in pure culture.

Comparable evidence for the importance of vessel tylosis as a defence mechanism is also available for vascular wilts of tomato. Blackhurst & Wood (1963) reported that the percentage of vessels occluded by tyloses, 19 days after root inoculation with *Verticillium albo-atrum*, was higher in a wilt-resistant variety, Loran Blood (36%), than in a susceptible variety, Ailsa Craig (23%). In a similar comparison between two varieties susceptible and resistant, respectively, to *Fusarium oxysporum* f. *lycopersici*, Beckman (1966) found a strong inverse correlation between density of vascular mycelium and the speed and completeness of vascular occlusion in xylem-inoculated plants. In the resistant variety, wilt symptoms were slight or absent, vascular discoloration extended only 1-2 cm above the injection point and further progress of the infection was halted by complete vascular occlusion. In the susceptible variety, wilt symptoms were fully expressed, vascular discoloration was extensive and tylose formation was sparse, with complete vessel occlusion occurring only at infrequent intervals. These findings confirm similar but earlier ones by Snyder, Baker & Hansen (1946), Scheffer & Walker (1954), and Scheffer (1957), who reported equally striking differences in development of Fusarium wilt in xylem-inoculated plants of susceptible and resistant varieties, respectively, of tomato. A second conclusion to be derived from all these experiments is that, because the two tomato varieties were xylem-inoculated, the difference in disease development following inoculation could have been mediated *only* by a difference in xylem defence-response between the two cultivars. So for tomato, as for

69

banana, that part of Talboys's hypothesis disclaiming a host-specific differential in the xylem defence-response must now be discarded as invalid.

It is only fair to Talboys to add, however, that this slight modification of his general hypothesis detracts very little from its great value in clarifying our understanding of how wilt diseases develop. Indeed, Talboys (1964) had explicitly foreseen the need for some modifications of this type to fit particular cases, because he has remarked in his concluding statement that 'the concept in no way excludes the possibility of additional or alternative determinants operating within the vascular system...for example, there may be variations in the threshold level of infection at which vascular occlusion is suppressed'. In making this proviso, Talboys had foreseen the possibility of host-specific variation in the xylem defence-response, evidence for which has been given above.

EFFECTS OF ENVIRONMENT AND SOIL NUTRIENT SUPPLY

Prior to the cortical invasion of young roots, there is opportunity for soil conditions to exert a direct effect on the pre-penetration stage of infection by vascular pathogens, beginning with germination of infective propagules in response to a nutrient stimulus from root exudates (Ch. 7). But by comparison with the indirect effect that soil conditions must exert upon the fungus via the physiology of the host plant during its progress through the tissues, this direct effect is probably small. There is, indeed, much evidence for the view that environmental conditions known to reduce the incidence of vascular wilt diseases do so more by retarding or inhibiting the internal progress of infection than by any direct effect on the pre-penetration phase.

More is known about the effect of soil and other environmental conditions on development of vascular wilts due to formae speciales of *Fusarium oxysporum* than has been established for wilts caused by *Verticillium* species and other vascular pathogens. In an earlier review (Garrett, 1956a, p. 223), I have tabulated and discussed evidence for the view that wilt diseases caused by *F. oxysporum* are favoured by: (1) somewhat high temperatures

(> 25° C); (2) soil acidity; (3) high available nitrogen content of soil; (4) sometimes by a deficiency in soil potassium supply.

In many of the extensive banana-growing areas of the Caribbean region, Panama disease has become a limiting factor in cultivation

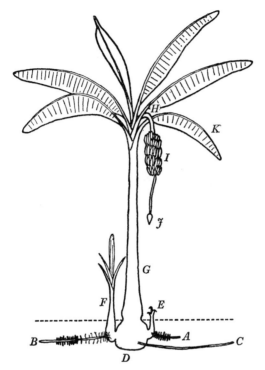

6. Growth habit of a Gros Michel banana plant. *A*, young root with turgid rootlets; *B*, older root with some withered rootlets; *C*, long mature root with no rootlets remaining at proximal end; *D*, rhizome; *E*, young bud or 'sucker'; *F*, older bud with small shoot; *G*, pseudostem, consisting of overlapping leaf-bases; *H*, flowering stem which during development grew up through the pseudostem; *I*, fruit bunch; *J*, male inflorescence; *K*, leaf. (After J. Rishbeth, 1955.)

of the popular but disease-susceptible Gros Michel; the large banana industry of Jamaica was saved by a change to widespread planting of the resistant Lacatan banana in the early nineteen-fifties. It is not surprising, therefore, that much is known about the effect of soil conditions on the incidence of Panama disease from field surveys and from glasshouse and field experiments. Amongst earlier investigators of this disease, Wardlaw (1941) travelled very widely over the Caribbean region, and concluded that Panama disease was favoured above all by acid soils, and that after this, a light soil texture was of next most importance as a factor in disease proneness. Conversely, clay soils and slightly alkaline soils appeared to retard development and spread of the disease, though Wardlaw was careful to add that this situation was not maintained indefinitely under continuous banana cultivation; this is merely an example of the ultimately disastrous development of a disease on the 'wrong' type of soil that I have already discussed in Ch. 1 and attributed to a gradual increase of pathogen population in the soil under intensive cultivation of a single crop. Wardlaw's conclusions have in general been confirmed by the extensive field surveys and experiments of Rishbeth (1955, 1957, 1960) and Rishbeth & Naylor (1957) over a period of 3 years in Jamaica. From a preliminary study of the course of infection, Rishbeth (1955) concluded that initial infection occurred through the rootlets and that access to the vascular strands of the long roots was obtained by way of the vascular strands of infected rootlets. Plant pathologists familiar with the root systems of cereals and grasses will find no more difficulty in understanding the root system of the banana than I did myself, but for the sake of others I have reproduced here Rishbeth's elegant drawing of the banana plant (Fig. 6); also reproduced is his illustration of infection of long roots from their rootlets (Fig. 7).

Rishbeth further observed that various fusaria other than *F. oxysporum* f. *cubense* quite frequently infected both rootlets and the long roots, but even in the susceptible Gros Michel they usually failed to make any substantial progress into the rhizome. To distinguish *F. oxysporum* f. *cubense* from non-pathogenic forms of *F. oxysporum* requires a pathogenicity test against young banana plants in pots and is therefore very laborious; it is an essential procedure, however, because Rishbeth and various other workers have upheld Hansford's (1926) contention that it is impossible to identify *F. oxysporum* f. *cubense* from colonies growing on the agar plate, and so the results of all earlier surveys of the distribution of

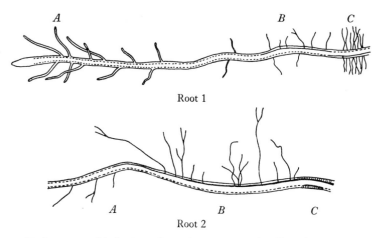

Root 1

Root 2

7. Early stages of infection of two banana roots by *Fusarium oxysporum* f. *cubense*; vascular strands are indicated by dotted lines.

Region	Extent (cm)	Characteristics of region
Root 1:		
A	10	Rapidly growing apical region with white, turgid rootlets, clustered near apex; healthy.
B	7	Central region with scattered, withered rootlets; limited vascular infection arising from four rootlets.
C	3	Proximal region with densely clustered, withered rootlets; vascular infection arising from one rootlet and entering rhizome.
Root 2:		
A	3	Cortex healthy, no infection through rootlets; vascular strand infected from region B, its discoloured zone narrow and yellow to orange in colour.
B	4	Cortex healthy; much infection through rootlets; vascular strand with a wide discoloured zone, very dark red in colour.
C	1	Cortex with blackish lesions (indicated by dots); vascular infection passing into rhizome.

(After J. Rishbeth, 1955.)

this forma specialis in banana soils that were made without the precaution of pathogenicity testing are open to serious question. By pathogenicity testing in this way, Rishbeth determined that 91 % of *F. oxysporum* isolates from the rhizome and pseudostem belonged to f. *cubense*, whereas only 29 % of those from roots and rootlets proved to be the wilt pathogen. Rishbeth therefore concluded that a barrier to the further development of infection occurred at the junction between roots and rhizome; even in Gros Michel, the pathogen was found to enter the rhizome only from some 5% of infected roots. Heavy soil fertilization with nitrogen increased the proportion of rhizome infections resulting from infected roots (Fig. 8).

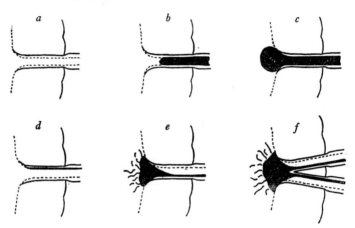

8. Infection of banana root bases by *Fusarium oxysporum* f. *cubense* and other fungi; vascular strands are indicated by dotted lines.

Diagram	Condition	Score for count of rhizome infection
a	Healthy	o
b	Vascular infection sealed off at root base, usually dark red in colour	o
c	Limited 'pocket' rot of rhizome caused by *Cylindrocarpon* sp.; no progressive vascular infection. Vascular strand of root usually yellow; rot dull reddish-purple	o
d	Vascular infection probably about to enter rhizome, pale yellow to orange	o
e	Vascular infection, caused by *Fusarium oxysporum* f. *cubense*, shortly after entering rhizome, orange-yellow, later dark brown or purple	I
f	As last, but rhizome infected almost simultaneously from two roots	2

(After J. Rishbeth, 1955.)

From his field surveys of the Jamaican banana areas, Rishbeth (1957) stated that healthy old plantations of Gros Michel were virtually confined to alkaline soils of medium to heavy texture, well drained and aerated, rich in available potash and not subject to heavy flooding. Conversely, he found the incidence of Panama disease to be widespread and severe on light-textured acid soils of low nutrient status, especially if drainage was poor. He suggested that an important effect of heavy rainfall is to wash spores of *F. oxysporum* f. *cubense* from the soil surface downwards into contact with the banana root system; there is indeed general agreement between earlier observers that the rapid extension of soil infestation by this pathogen from 1912 onwards in Jamaica has been due to sheet flooding after heavy rainfall, especially on slowly draining soils (F. E. V. Smith, 1936).

In a series of manurial experiments on five sites in Jamaica, all quite heavily infested with the wilt pathogen, Rishbeth & Naylor (1957) failed to find any significant reduction in percentage wilted plants following applications of lime, phosphate or potash, whether separately or in combination; the only significant result from these experiments was a demonstration of the unwelcome effect of nitrogenous fertilizers in aggravating incidence of the wilt. This result was disappointing in view of the correlation found by Rishbeth (1957) from disease surveys between comparatively healthy banana plantations and a high soil content of available phosphate and potash. Small plantations, up to 5 acres in extent, on sites of former sugar factories often formed islands of healthy bananas, surrounded by land abandoned to banana cultivation after devastation by Panama disease; the soil of such sites had an exceptionally high status in available phosphate and potash, due to dumping or burning of sugar cane trash, remaining after extraction of sugar in the factory, over a long period of years. Such observations on soils resistant to the development of Panama disease, even under long continued banana cultivation, suggested to Rishbeth (1960) that the difficulty in imitating these presumed effects of phosphate and potash by application of fertilizers consisted in getting the banana plant to take up *enough* of these nutrients from the soil. In a glasshouse experiment conducted in Cambridge after his return from Jamaica, he grew banana plants in a series of inoculated soils so graded with respect to texture and moisture régime that soil aeration varied from poor through moderate up to good; the results showed that minimum development of wilt occurred in the best aerated soil, and was correlated with the maximum uptake of potash. The results of this experiment thus confirm Rishbeth's (1957) earlier conclusion from his field surveys that banana plantations on well drained and aerated soils with a high potash status tend to remain comparatively free from Panama disease. In its degree of physiologically conditioned resistance to the development of wilt, the banana plant, like others, thus reflects the interaction of many soil factors, which it integrates in its growth.

The effect of nitrogen in promoting the development of vascular wilt diseases seems to be a very general one, as it has been widely reported for wilts caused by species of *Verticillium*, as well as for those due to formae speciales of *Fusarium oxysporum*. Spencer & McNew (1938) found that development of a vascular wilt of sweet corn caused by *Xanthomonas stewarti* (E. F. Smith) Dowson was promoted by excess nitrogen and tended to be suppressed by excess potassium in sand culture experiments; from their trials of xylem sap from plants grown under different nitrogen régimes as a growth medium for *X. stewarti*, Spencer & McNew suggested that the observed effect of nitrogen in promoting wilt development might be explained by a direct nutritional effect on growth of the bacterial pathogen in the xylem tracts. A hypothesis to explain the effect of potassium deficiency in aggravating development of vascular wilts was proposed by Shear & Wingard (1944), based on findings by Shear (1943) and others that a deficiency of potassium in the soil solution led to abnormally high levels of nitrogen and phosphorus in the xylem sap. These high levels of nitrogen and phosphorus were attributed to a block in the synthesis of organic derivatives, caused by the deficiency of potassium. At the time that Shear & Wingard put forward this hypothesis, nitrogen was generally thought to be translocated in the xylem in the form of nitrate. This assumption is now known to be incorrect, as in many plants the nitrogen in the xylem sap is made up chiefly of amino acids and amides. This does not necessarily invalidate Shear & Wingard's hypothesis; it is conceivable that in plants growing under poor light conditions or under the stress of potassium or other mineral deficiency, a low level of photosynthesis could result in absorption by the roots of nitrogen surplus to carbohydrate production, and that this surplus nitrogen would then be transported up through the xylem; an increase in *total* nitrogen content of the xylem sap in potassium-deficient plants could only occur, however, if an appreciable fraction of the organically fixed nitrogen in normal plants stays behind in the root system and is stored there. Quite apart from its implications for Shear & Wingard's hypothesis, this fairly recent work on the way in which nitrogen from the root system is transported in the xylem must be of direct interest to workers on these vascular wilt diseases. Bollard (1960), who has made extensive studies of the composition of xylem sap in apple and other rosaceous fruit trees, has stated that, at least in trees growing under good conditions, the nitrogen fraction of the sap is made up of organic nitrogen compounds, chiefly amino acids and amides; nitrate appears to be reduced and converted into these organic forms within the feeding rootlets of the tree. Bollard has extended this examination of the xylem sap to a wider range of woody plant species, and from this it appears that nitrogen is generally transported up the xylem in organic forms, though in some species it is accompanied by appreciable concentrations of nitrate. Some observations made during studies of vascular wilt diseases are compatible with Bollard's conclusions. Thus Zeevaart (1955) found no sugars, but did detect aspartic acid at a concentration of approx. 0·02 % in the xylem sap of lupins; in plants infected by *Fusarium oxysporum* Fr. f. *lupini* Snyder & Hansen, the concentration of aspartic acid in the sap was much lower. Xylem sap was collected from five tree species in summer by Kessler (1966) and found to constitute a satisfactory medium for growth and sporulation by four vascular wilt fungi; the saps contained no sugar but did contain a mixture of amino acids and/or amides, which provided a carbon as well as a nitrogen source for growth of the four fungi.

Earlier in this discussion of the effect of environmental conditions on vascular wilts, I pointed out that such conditions must chiefly affect the pathogen within the tissues via their influence on the physiology of its host. Such a realization led Foster & Walker (1947) to find out if they could reproduce the effect of environmental conditions upon tomato wilt due to *Fusarium oxysporum* f. *lycopersici* by growing healthy young tomato plants under various environmental combinations for a 'conditioning period' *before* inoculation with the pathogen; after inoculation, all sets of plants were held under the same, uniform environmental conditions, selected as optimum for wilt development. Environmental factors investigated in this way were termed 'predisposing factors'. Certain conditions of temperature, light, soil water content and pH that were already known to favour wilt development also acted in the same direction when acting on the plant only as predisposing factors, and not subsequent to the act of inoculation. But effects of nitrogen and potassium manuring as predisposing factors went in the opposite direction to that expected; plants previously grown with high nitrogen or with low potassium were actually less susceptible to subsequent wilt development than those grown in the opposite way.

It is difficult to understand why nitrogen and potassium acting only as predisposing factors in these experiments by Foster & Walker should have behaved in a way opposite to that when acting as environmental factors before *and* after infection. This discrepancy has not appeared in the results of Al-Shukri (1969), who employed the technique devised by Foster & Walker for a comparable study of four vascular wilt pathogens of cotton: *Verticillium albo-atrum*, *V. dahliae*, *V. tricorpus* and *Fusarium oxysporum* Fr. f. *vasinfectum* (Atkins.) Snyder & Hansen. Wilt incidence was recorded as percentage plants wilted by the end of each experiment, and wilt severity as the reciprocal of the mean incubation period to first appearance of wilt symptoms in each treatment; the second estimate provided much the more sensitive index to the effects of the treatments. For all four species of pathogen, a high level of nitrogen (as hoof and horn fertilizer) predisposed plants to development of the wilt disease; a deficiency of potassium showed a similar effect, though this was not always statistically significant. Soil acidity predisposed plants to development of all four wilt diseases, which was of particular interest in view of the general correlation between soil acidity and liability of many crop plants to wilt diseases caused by formae speciales of *Fusarium oxysporum*. All of Al-Shukri's findings are thus compatible with the supposition that effects of soil conditions, including mineral nutrients, upon development of vascular wilt diseases operate mainly through changes induced in host-plant physiology, and that effects upon the pathogens in the soil and during the act of root infection are subsidiary to this.

4   Specialized parasites: ectotrophic root-infecting fungi

This group of fungi includes nearly all the specialized root pathogens apart from the vascular wilt fungi discussed in the preceding chapter. In the ectotrophic pathogens, infection of the root tissues is typically preceded by the superficial extension of mycelium over the root-surface; the interval between the advancing margin of epiphytic mycelium and the infection that follows it can vary from a millimetre or so with some pathogens of herbaceous plants to several metres with some fungi infecting roots of tree species. As I described in Ch. 2 for *Rhizoctonia solani*, root infection by unspecialized fungal pathogens may also be preceded by superficial mycelial growth and attachment to the root surface. But because such unspecialized pathogens are usually limited by mature tissue-resistance to immature, senescent, wounded or unhealthy parts of the root system, this superficial spread of mycelium does not continue indefinitely; sooner or later, it is halted by host resistance. The specialized pathogens, on the other hand, are not thus halted by the development of effective defence mechanisms in older parts of the root system; provided always that soil conditions are favourable for ectotrophic infection, an essential proviso, they can continue to spread from the original foci of infection, perhaps on outlying parts of the root system, up to the root collar and stem base. The continuous, ectotrophic infection-habit of these specialized pathogens is thus no more and no less than the outward and visible sign of an infectivity that is adequate to overcome host resistance to infection throughout all parts of the root system.

There is an obvious superficial resemblance between the infection-habit of these specialized pathogens and that of the ectotrophic mycorrhizal fungi forming symbiotic associations with many species of forest tree; it was this resemblance that led me to propose the name 'ectotrophic' for these specialized pathogenic fungi as well (Garrett, 1956a, Ch. IV). The use of this term 'ectotrophic' does indeed provide a convenient and unequivocal morphological description of the infection-habit of the pathogens, but it cannot be taken to imply any relationship with the ectotrophic mycorrhizal fungi that is more profound than a similarity in morphology of the infection-habit. It can be suggested that the ectotrophic mycorrhizal fungi have evolved from the specialized root pathogens, according to the general postulates about the evolution of parasitism that I discussed in Ch. 1. No less plausible, however, was Harley's (1948) suggestion that ectotrophic mycorrhizal fungi may have evolved from saprophytic root-surface fungi. Later in this chapter, I shall present much evidence for the conclusion that the ectotrophic infection-habit of the pathogens functions as an efficient mechanism for overcoming host resistance. Such a view is not appropriate for the ectotrophic infection-habit of the mycorrhizal fungi; in this apparently harmonious host–parasite relationship, there is no evidence that the fungus provokes any significant degree of host resistance, at least so far as the limited degree of infection that actually develops is concerned. Harley (1969) has described the morphogenesis and behaviour of various types of ectotrophic mycorrhizal association in critical detail. But a single example will suffice here to make my point. In his work on the ectotrophic mycorrhizal association of Scots pine, Robertson (1954) demonstrated that although infection of the short roots is truly ectotrophic (i.e. they are enveloped by the usual sheath of fungal mycelium), that of the long roots proceeds *internally*, as an extension of the Hartig net intercellularly in the cortex (Fig. 9).

From Robertson's evidence, it is clear that the ectotrophic infection-habit is not essential for successful extension of infection along the young long roots of Scots pine. I shall discuss later parallel situations with ectotrophic fungal pathogens, in which a decline of host resistance to a low level permits a purely internal extension of infection along roots of much impaired vigour. This suggests that the ectotrophic infection-habit is not imposed upon ectotrophic mycorrhizal fungi, as it seems to be on ectotrophic pathogens, by the occurrence of a strong provoked resistance to infection. It thus seems most probable that the external fungal sheath of ectotrophic mycorrhizas has developed purely as an absorbing organ for the taking-up of nutrient salts from the soil. This seems to have been a roundabout way for a plant pathologist to reach a conclusion that most mycorrhizal workers have probably always taken for granted.

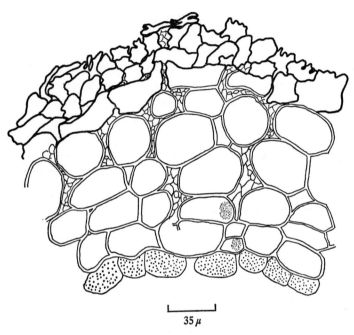

35 μ

9. Transverse section of cortex of mycorrhizal long root of *Pinus sylvestris* approx. 8 mm behind the growing apex. Note Hartig net of mycelium between cortical cells. (After N. F. Robertson, 1954.)

INFECTIVE MECHANISM OF THE ECTOTROPHIC INFECTION-HABIT

The essential features of the ectotrophic infection-habit amongst these specialized pathogens are well shown by *Ophiobolus graminis*, restricted to the Gramineae and causing the take-all disease of cultivated cereals. The fungus progresses at first epiphytically over the root system as a sparse network of dark-coloured runner hyphae, from which hyaline branches quickly penetrate and infect the underlying cortex and eventually the vascular cylinder of the roots (Fig. 10).

In my early work on the ectotrophic infection-habit of *O. graminis*, I noted briefly two observations, the full significance of which I did not realize at that time. First, I observed that as the

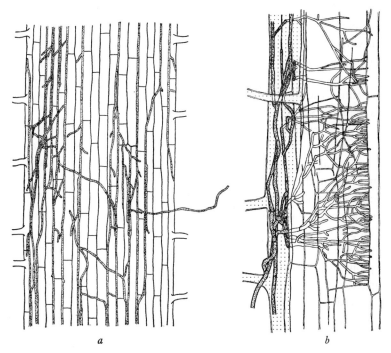

a   *b*

10. Ectotrophic infection of wheat seedling root by *Ophiobolus graminis*: (*a*) in surface view; (*b*) in longitudinal section. (After drawing made by G. Samuel, in Garrett, 1934.)

fungus travelled down a seminal root from the point of inoculation below the germinating seed, it often abandoned the ectotrophic habit on the distal part of the root and the runner hyphae thereafter travelled intercellularly one or two cells deep within the cortex (Garrett, 1934). I attributed this to an impairment in resistance to infection of the distal part of the root, resulting from its virtual disconnection from the rest of the plant through destructive infection of the proximal part. The significance of this observation will become more apparent later, though I myself did not appreciate its full implications at that time. Secondly, I noted that if ectotrophic infection of a root was inhibited by sufficiently unfavourable conditions, such as too extreme a degree of soil acidity or too poor a degree of soil aeration, then *O. graminis* also failed to extend appreciably further by purely internal infection of the root cortex. My comment on this was 'It is not yet clear why this is so, since growth of the infection hyphae transversely across the root appears to proceed without any difficulty' (Garrett, 1936). Subsequent experiments with other ectotrophic root parasites have now explained the significance of my early observations on *O. graminis*, as we shall shortly see.

Our understanding of the mechanism of the ectotrophic infection-habit has evolved gradually from the earliest attempts at a teleological explanation in terms of efficiency up to the present time when the teleological hypothesis has been confirmed and replaced by a mechanistic explanation that is compatible with all the known facts. First in the field were the lively speculations of Napper (1934) at the Rubber Research Institute of Malaya. He suggested that infection could progress more rapidly from an epiphytic mycelium that sent in infection hyphae or hyphal aggregates (i.e. rhizomorphs or mycelial strands) at intervals than it could by purely internal progress through the root tissues, in which invasion would be retarded both by the passive mechanical resistance of cell walls and by active defence-responses of the living tissues. In my earlier discussion of this general problem (Garrett, 1956a, Ch. IV), I rehearsed the evidence on which such a teleological hypothesis could be founded. I postulated that the ectotrophic infection-habit provided a mechanism whereby the parasite could initiate a series of infections in comparatively rapid succession along the length of the root, thus diluting and eventually overwhelming the active defence-response of the root to infection. This hypothesis thus clearly implied synergism between a series of successive infections in overcoming active tissue-resistance of the root. At that time, this explanation rested largely upon evidence collected by Rishbeth (1950, 1951b) upon the ectotrophic infection-habit of *Fomes annosus* upon the roots of *Pinus sylvestris* (Scots

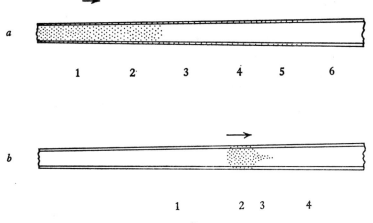

11. Diagram representing two modes of infection of pine roots by *Fomes annosus*.

(*a*) *Infected root from alkaline soil*

Region	Extent (cm)	Characteristics of region
1	> 15	Little resin exudation; wood dry; *F. annosus* hyphae abundant.
2	10	Marked resin exudation; wood resinous; hyphae less abundant.
3	10	Marked resin exudation; wood resinous; hyphae in bark only.
4	17	Local resin exudation; wood not resinous; hyphae in bark only.
5	25	Bark scales whitish, peeling; a few superficial hyphae present.
6	—	Root normal, bark scales intact.

(*b*) *Infected root from acid soil*

1	> 15	Bark split by resin; wood resinous; secondary fungi present only.
2	4	Marked resin exudation; wood resinous; *F. annosus* hyphae in bark and wood.
3	2	Slight resin exudation; wood resinous; hyphae in centre of wood only.
4	—	Root normal.

▦ Indicates region where *F. annosus* is present.
→ Indicates direction of invasion by *F. annosus*.

(After J. Rishbeth, 1951*b*.)

pine) and *P. nigra* var. *calabrica* (Corsican pine). In alkaline soils, this fungus grows freely on the outside of pine roots, and the sheet-like mycelial aggregates amongst the bark scales can some-times be found as much as half a metre ahead of infection in the wood cylinder. In acid soils (approx. pH 4·5), on the other hand, unfavourable soil conditions seem much to delay, and sometimes to suppress altogether, advance of the epiphytic mycelium, with the result that a slow progress of infection in the wood cylinder is sometimes ahead of epiphytic mycelium on the bark (Fig. 11).

Rishbeth's work with *F. annosus* thus laid the foundations for an understanding of the ectotrophic infection-habit; the critical experiment that has led to a satisfactory mechanistic explanation was carried out and published by one of Rishbeth's research associates, Wallis (1961). Wallis's experiments went very directly to the point at issue, i.e. whether the ectotrophic infection-habit is imposed upon the fungus by the active defence-response of the living root tissues. Wallis therefore made use of the earlier finding by R. Leach (1939) in East Africa that severing a tree root from the parent tree greatly reduces its power of active resistance to infec-tion and spread by *Armillaria mellea*. In Wallis's experiments, roots of 20-year-old Scots pine trees growing on alkaline (pH 7·0–7·5) and acid (pH 3·8–4·3) soils, respectively, lying at a depth of 15–25 cm, were exposed and brushed free of soil. At 70–90 cm from the base of the tree, a root section 8–12 cm long was cut out and replaced by a woody inoculum block permeated by mycelium of *F. annosus*. Twelve months after inoculation, both external spread of *F. annosus* over the bark and the extent of internal infection of the wood cylinder were measured for all the treatment-series; results for dominant trees in normal vigour have been taken from Wallis's Tables IV and V and are summarized here in Table 3.

Wallis's results as summarized in Table 3 are so clear-cut as to need little comment. Behaviour of *F. annosus* on the proximal root portions, still connected with the parent tree and thus still in possession of full vigour and a full active defence-response to infection, is typical of that earlier observed by Rishbeth; ecto-trophic advance of the fungus on the outside of the roots in

Table 3. *Extent of external mycelial spread and of internal infection of wood cylinder by* Fomes annosus *on roots of Scots pine (measurements in cm)*

(From Wallis, 1961)

	Acid soil		Alkaline soil	
	Outer bark scales	Wood	Outer bark scales	Wood
Proximal root portions (connected)	5	3	15	0·7
Distal root portions (disconnected)	15	48	56	75

alkaline soils is much ahead of that in acid soils, and it is only a matter of time before internal infection of the wood cylinder catches up with it. In the distal root portions, disconnection from the parent tree has evidently reduced the active defence-response to a level at which *F. annosus* has been able to spread more rapidly *within the wood cylinder* than it could do outside the root, even under the more favourable conditions for external spread, i.e. in the alkaline soil.

From the results of Wallis's experiments, it has thus become clear that the ectotrophic infection-habit is imposed upon these specialized root pathogens by the active defence-response of roots in possession of their full vigour. If no active defence-response is provoked, then internal infection can proceed without obvious hindrance, as I have already suggested in explanation of Robertson's (1954) observation on the internal extension of the Hartig net in the long roots of Scots pine (Fig. 9). The level of active defence-response of living root tissues varies to some extent in different host–parasite combinations. As I have noted earlier, it seems to be sufficient to prevent a purely internal extension of infection by *Ophiobolus graminis* within wheat roots. The same conclusion apparently holds for infection of roots of the rubber tree by *Fomes lignosus*; thus John (1958) demonstrated that infection of rubber roots by this fungus could be completely halted merely by removing the epiphytic rhizomorphs at weekly intervals whilst leaving the original inoculum undisturbed. In contrast to these two pathogens, *Fomes annosus* can sometimes progress slowly along a pine root by internal infection of the wood cylinder, even though such internal infections are quite often brought to a halt by the active defence-response of the living host-tissues; the nature of this defence-response of pine roots will shortly come up for discussion.

Following upon this recent work, it has now become possible to explain the mechanism of the ectotrophic habit on the assumption that the active defence-response of the host root is diluted and finally overcome by synergism between the effects of separate infections arising in succession from the externally travelling mycelium on the outside of the root. I can best illustrate this by an analogy with the epidemic development of the chocolate spot disease of field beans (*Vicia faba*), caused by species of *Botrytis* and chiefly by *B. fabae*. Wilson (1937) successfully elucidated the epidemiology of this disease by distinguishing between its *non-aggressive* and *aggressive* phases. The non-aggressive phase consists of scattered leaf-spots, or primary lesions, caused by single-spore infections (Wastie, 1962); if the primary lesions are scattered thinly on the leaf, their further expansion is arrested by the host defence-response, and they remain static. But as Wilson demonstrated, one essential condition for development of the aggressive phase appears to be a degree of crowding of the primary lesions that exceeds a certain critical value; above this critical value, synergism between the effects of the individual infections operates in such a way as to overcome host resistance. The fungus is thus enabled to break out of the primary lesions and to initiate the aggressive phase of rapid spread through leaf and stem tissues, which continues unless and until interrupted by a spell of warm, dry weather. At the Cambridge Botany School, we have confirmed and extended Wilson's findings in a yearly demonstration experiment for students set up by Mr W. J. Bean, which has continued for nearly 20 years. In Fig. 12, I have provided a diagrammatic representation showing in parallel what I conceive to be the sequence of events following upon: (*a*) a single isolated infection; (*b*) a multiple series of infections, either contemporaneous or shortly consecutive, first for the chocolate spot disease of beans and secondly for a typical, ectotrophic root-infecting fungal pathogen.

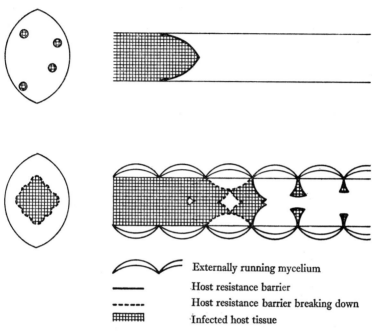

Externally running mycelium
Host resistance barrier
Host resistance barrier breaking down
Infected host tissue

12. Diagram illustrating possible mechanism whereby the ectotrophic infection habit enables a pathogenic root-infecting fungus to overcome host resistance. Compare this with a similar synergistic effect of multiple infections by a leaf-infecting fungal pathogen, which enables its invasion of the leaf to become progressive.

NON-ECTOTROPHIC INFECTIONS

Under this heading, it is appropriate to describe the behaviour of *Fomes annosus* on some species of conifer other than those of *Pinus*, as elucidated by Rishbeth (1951c). In these other conifers, such as Douglas fir (*Pseudotsuga taxifolia*), larch (*Larix decidua*) and spruce (*Picea abies*), resistance to infection of the inner wood tissues is lower than that in species of pine, and so infection follows a very different course. In these conifers, as in pines, exposure of very young trees to heavy or multiple infections can result in outright killing; in older trees, however, infection leads to development of butt rot, which can cause much heavier losses of marketable timber than does the more dramatic tree-killing as manifested in pine plantations. In trees, some 20–30 years old, of Douglas fir, larch or spruce, the following infection sequence has been described by Rishbeth. The youngest roots are generally the first to become infected, usually by contact with infected stump roots resulting from basidiospore infection by *F. annosus* of the tree stump surfaces, exposed by tree felling during a plantation thinning (Rishbeth, 1951 *a*). In these small roots, there is little resin

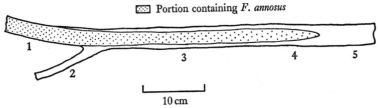

13. Infection of larch root by *Fomes annosus*, of the type that leads to eventual development of butt rot. 1. Wood very rotted, yellow; bark, dead, brown; *F. annosus* hyphae abundant in both. 2. Small living root. 3. Most of the wood resinous, red-brown, hyphae abundant; outermost wood healthy; bark discoloured, bright red in parts, otherwise healthy, hyphae local on surface only. 4. As last, but smaller region of central wood with hyphae. 5. Whole root healthy. (After J. Rishbeth, 1951 *c*.)

response by the root to the developing infection, and *F. annosus* grows rapidly in both bark and wood until it reaches an older part of the root of some 2 cm in diameter. Thereafter, resistance of the central wood is lower than that of the outer sapwood, and so the fungus advances through an inner cylinder of wood, leaving the outer sapwood uninfected and still able to produce branch roots (Fig. 13).

As *F. annosus* approaches the root collar, or butt of the tree as it is sometimes called, the boundary between the central cylinder of infected wood and the outer uninfected wood is often marked by a conspicuous layer of resin (Fig. 14).

It is thus apparent that in species of conifer susceptible to the butt-rot type of infection, internal spread of *F. annosus* is made possible by the fact that the resistance of the inner cylinder of wood is sufficiently lower than that of the outer sapwood in these species, and is also lower than that of any part of the wood cylinder in pine species. The reasons for these differences have been elucidated in an investigation of outstanding interest by Gibbs (1968), working as one of Rishbeth's research associates. By direct inoculation of *F. annosus* into the wood cylinder of the roots, Gibbs was able to estimate the resistance of the wood in Douglas fir, Sitka spruce (*Picea sitchensis*), Scots pine, and Corsican pine, respectively; the

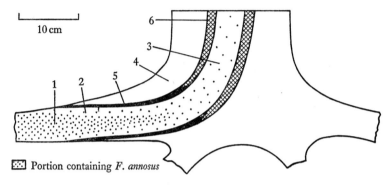

☷ Portion containing *F. annosus*

14. Diagrammatic section of lower butt of Douglas fir, showing incipient development of butt rot, with *Fomes annosus* entering from a lateral root. 1. Wood extensively rotted, yellow. 2. Wood, partially rotted, yellow-brown. 3. Wood with incipient rot, red-brown. 4. Wood healthy, white. 5. Resin layer. 6. Wood resinous, deeply stained. (After J. Rishbeth, 1951c.)

degree of resistance was estimated as the reciprocal of the distance travelled by *F. annosus* in the wood over a given period. Douglas fir showed least resistance, and the two pine species the most; the degree of resistance up the ascending series from Douglas fir to the pines was closely correlated with the complexity and activity of the resin canal system. This is nicely illustrated, as Gibbs has pointed out, by commercial use of these various genera of conifers for commercial production of resin; various species of *Pinus* are widely employed for resin tapping, those of *Picea* much more rarely, and those of *Larix* and *Pseudotsuga* scarcely at all. Gibbs further noted from the results of his inoculation experiment that in spruce and Douglas fir, *F. annosus* failed to infect the outer sapwood and was restricted to the inner wood, in which a high proportion of the resin ducts become atrophied. In other experiments, with Scots and Corsican pine, Gibbs recorded daily resin yields by tapping; resin yield, tree vigour and degree of resistance to *F. annosus* all varied in parallel under the conditions obtaining in these trials.

An analysis of the toxicity of resin (in this paper designated as 'oleoresin') to *F. annosus* and four timber bluestain fungi (*Ceratocystis* spp.) has been made by Cobb, Krstic, Zavarin & Barber (1968). They found that both crude oleoresin and nine components of its volatile turpentine fraction, purified by fractional distillation, reduced mycelial growth of *F. annosus* and of all four bluestain fungi by various degrees when the fungi were exposed to a vapour-saturated atmosphere; growth reductions were still greater when either oleoresin or one of its volatile components were incorporated in the growth medium. These nine turpentine components were identified; there were two monocyclic, four bicyclic and one open-chain terpene, and two alkanes. The most fungistatic of these components was the alkane *n*-heptane, which in a vapour-saturated atmosphere completely inhibited mycelial growth of both *F. annosus* and *C. pilulifera*.

MYCELIAL AGGREGATION INTO SHEETS, STRANDS AND RHIZOMORPHS

Ectotrophic pathogens that infect woody perennials, i.e. bushes and trees, typically show aggregation of the epiphytic mycelium into mycelial sheets or strands, or into rhizomorphs. They thus contrast with fungi infecting herbaceous species, such as *Ophiobolus graminis*, in which the epiphytic mycelium is not organized into hyphal aggregates but consists of separated individual hyphae. This distinction is not affected by the fact that *Armillaria mellea*, a typical ectotrophic parasite of tree root systems, will infect potato tubers if opportunity offers; were *A. mellea* limited to potato plants and other herbaceous species, it is unlikely that it would have evolved so powerful an offensive armament as the fungal rhizomorph. Evidence of three kinds suggests that these various forms of hyphal aggregation have evolved in response to the need for an inoculum potential adequate to overcome host resistance to infection by the roots of woody perennials, and particularly by those of the larger arborescent species. First, mycelial aggregates as organs of infection are not found in root-infecting fungi restricted to herbaceous hosts. Secondly, many experiments have now shown that there is a critical level of inoculum potential necessary for initiation of a progressive infection; the older, larger and more highly organized host roots demand a higher level of inoculum potential for establishment of a progressive infection than do younger, smaller and simpler roots. Thirdly, there is no evidence for the supposition that the older, cork-covered parts of tree root systems can be successfully infected either by spores or by individual hyphae in unorganized mycelium, though there is much evidence for the contrary view that infection can be accomplished only by mycelial aggregates growing out from a food base generating adequate inoculum potential.

MORPHOGENESIS AND BEHAVIOUR OF MYCELIAL STRANDS AND RHIZOMORPHS

Mature mycelial strands are superficially difficult to distinguish from rhizomorphs; both types of organ may reach a diameter varying from one to several millimetres, and a mature strand may have a surface as smooth as that of a rhizomorph. These two types of fungal organ, however, develop in very different ways and whether a particular organ is a strand or a rhizomorph can be determined only by a careful study of its morphogenesis. A mycelial strand is built up gradually, sometimes by the chance meeting of individual hyphae or minor strands; more commonly, however, strands develop acropetally by the ensheathing growth of branches of the parent leading hypha or hyphae. At least two types of strand morphogenesis have been described, and further variants on the general theme are likely to be discovered. Rhizomorphs, on the other hand, are morphogenetically similar to the roots of a higher plant; they grow from an apical meristem, though this meristem is of a type quite different from that found in any higher plant.

In consequence of this difference in morphogenesis, mycelial strands and rhizomorphs also differ in the way that their inoculum potential develops. The diameter of a rhizomorph attains its maximum value just behind the growing apex, and so the apical region possesses, to the maximum possible degree, 'instant inoculum potential'. A mycelial strand, in contrast, makes its first contact with a host surface with the arrival of its leading hypha (or associated leading hyphae); thereafter, both strand diameter and inoculum potential increase acropetally up the still developing strand, as further constituent hyphae grow up towards the strand apex from the rear.

Mycelial strands

The morphogenesis of the mycelial strands produced by *Helicobasidium purpureum*, causing the violet root rot disease not only of sugar beet but also of various woody perennial species, has been analysed by Valder (1958). Strands develop from a food base lying in or on the soil when environmental conditions are suitable for mycelial growth (Garrett, 1946). The food base employed by Valder in this study was a disc of nutrient agar colonized by *H. purpureum*, but under natural conditions both infected roots and sclerotia provide food bases for strand development. From such a food base, there first develops a sparse growth of robust, coloured hyphae over and through the soil; further hyphae growing out from the food base sooner or later encounter one of these leading hyphae and then follow it in associated growth. By continued accretion of these 'following' hyphae and also of small strands, each of the original leading hyphae may develop into a main strand, the base of which comes to resemble a river delta. Owing to this mode of morphogenesis, these strands are widest at the base and taper off towards the apex. Anastomoses occur between strands to give a network, especially near the food base. Branching of strands can occur, due to the branching of the original leading hypha, with the branch attracting some of the follower hyphae away from their parent hyphae. Valder further observed that coherence of the strands of *H. purpureum* was brought about in three ways: (1) by an interweaving growth of the associated main hyphae; (2) by the binding action of short branches of limited growth, sometimes hooked at the apex; (3) by anastomoses between hyphae (Fig. 15).

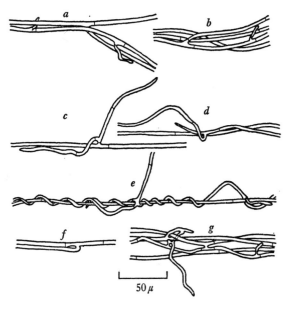

15. Morphogenesis of mycelial strands in *Helicobasidium purpureum*. The general direction of hyphal growth is from left to right. Note interweaving growth of main hyphae, binding action of short side-branches of limited growth (*a, b, g*), and the locking action of hyphal anastomosis (*g*). (After P. G. Valder, 1958.)

A second and commoner type of strand morphogenesis is exemplified by that of *Phymatotrichum omnivorum* (Shear) Duggar, which causes Texas root rot of cotton and other dicotyledonous species, including a number of woody perennials. Rogers & Watkins (1938) have described how the strands of this fungus develop around single, exceptionally wide hyphae, which become ensheathed by their own branch hyphae. By subsequent adjustment of position and septation in the ensheathing branch hyphae, a compact cortex of several layers in thickness is eventually formed around the single large hypha at the centre of each strand (Fig. 16).

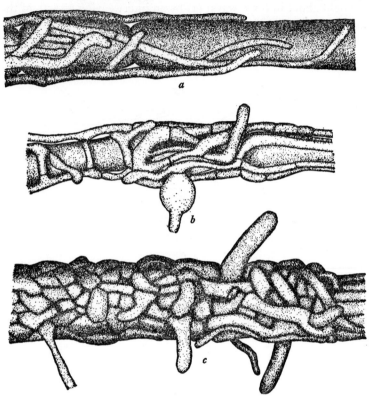

16. Morphogenesis of mycelial strands in *Phymatotrichum omnivorum*, (*a*) Small hyphae beginning to grow over the surface of the wide, central, leading hypha; (*b*) central hypha surrounded by a loose network of small hyphae; (*c*) deposition of the second hyphal layer. (After C. H. Rogers & G. M. Watkins, 1938.)

A type of strand morphogenesis very similar to that in *P. omnivorum* has been described for two well known saprophytes: *Merulius lacrymans* Fr., causing dry rot of house timber (Falck, 1912; G. M. Butler, 1958), and *Agaricus bisporus* (J. Lange) Pilát, the cultivated mushroom (Mathew, 1961).

Mycelial strands are typically the form in which a strand-producing fungus migrates outwards from a food base through the mineral matrix of the soil, which approximates to a non-nutrient medium, and at first epiphytically over root surfaces. When growing on a nutrient medium, such fungi either do not produce distinguishable strands, except amongst the aerial mycelium out of contact with the substrate, or they produce them only during the later stages of substrate colonization. In earlier accounts of strand morphogenesis (Garrett, 1960*a*, 1963), I had postulated an attraction between hyphae of the same mycelium or species as responsible for strand morphogenesis, and had suggested that this attraction might be similar to, or even identical with, that responsible for anastomoses between hyphae. Absence of strands, or delayed production, on nutrient media had then to be explained by a repulsion between hyphae in such a situation that was strong enough to override the postulated attraction. Causes of repulsion were not far to seek; on a nutrient medium every hypha acts as a diffusion sink for nutrients and as a source of growth products, some of which are likely to become fungistatic above a certain concentration. The consequences of these two processes were thought to cause hyphae to keep as far away as possible from one another on a nutrient medium, and thus to favour a dispersed type of growth.

Very recently, one of my research associates, Dr Sarah Day (1969), carried out an experiment with *Merulius lacrymans* to test this hypothesis; she argued that if it were correct, then *M. lacrymans* should form strands in a liquid nutrient medium that was renewed sufficiently rapidly by a drip feed device to prevent diffusion shells, either of nutrients or of fungal growth products, forming around individual hyphae. The results of this experiment were the reverse of those expected; mycelial strands formed in the still but not in the replacement cultures. From a consideration of these results and of many observations by earlier workers, Day was able to formulate a new hypothesis that seems to account satisfactorily for all the known facts. This hypothesis postulates that the leading hyphae, around which strands are formed, exude nutrients when they are translocating in much the same way as do young roots, and that it is this nutrient exudation that causes the chemotropic response of associated growth in the branch hyphae.

An indication of nutrient exudation by hyphae of *Rhizoctonia solani* in soil was earlier obtained by Thornton (1953), who observed these hyphae to be invested by actinomycete filaments. Evidence for nutrient exudation from fungal spores lying in soil will be presented in Ch. 7. But Day herself has obtained direct evidence of nutrient exudation by translocating hyphae of *M. lacrymans* from the results of experiments with radioactive isotopes.

Day's hypothesis thus provides an explanation for the regular occurrence of strand formation by strand-forming fungi when they are growing from a food base through natural soil or through any other non-nutrient media, such as those employed by G. M. Butler (1957) for *M. lacrymans*. The delay in stranding when a fungal colony is growing out over a nutrient medium can be explained by the masking of nutrient exudation from the leading hyphae through the initially high concentrations of nutrients in the medium; only when this initially high nutrient level has been sufficiently lowered by fungal growth will the masking effect be removed. So this hypothesis can account for the fact that strand formation by *M. lacrymans*, and by some other fungi too, does occur on nutrient media in due course, though its intensity varies with nutrient concentration and balance, as Day has shown.

An objection that can be raised against this hypothesis is that both strand formation and hyphal anastomosis occur only between mycelia of the same species, whereas the attraction provided by hyphal exudation is unlikely to be so specific, or, indeed, to have any significant degree of specificity at all. The answer to this objection is that the hyphae of a growing fungus are already loosely associated in their colonial growth, and that hyphal exudation needs only to hold together as a strand those hyphae that either arise in juxtaposition, i.e. branches of a leading hypha, or else approach one another by chance, as in strand formation by *Helicobasidium purpureum*. The same argument also applies to the mechanism of anastomosis; a nutrient stimulus from hyphal exudation will serve to prolong a hyphal association brought about by mere chance in the first place. Indeed, hyphal association represents the first stage only in the development of anastomosis; as Flentje (1969) has described for *Rhizoctonia solani*, various

98

causes of failure are spaced in the time sequence from initial juxtaposition of hyphal growth up to the ultimate consummation of anastomosis between compatible hyphae.

Rhizomorphs of Armillaria mellea

The best known rhizomorph is that produced by *Armillaria mellea*. If we assume the mean diameter of the straight running, un-branched, constituent hyphae to be about $10\,\mu$, then we can calculate that a rhizomorph of 1 mm diameter will be made up of several thousand hyphae, even after allowing generously for inter-hyphal space and for the central open channel running down the mature part of the rhizomorph. Growth of the rhizomorph, like that of a root, occurs from an apical meristem, and this implies a mechanism for coordinating apical growth of the constituent hyphae (Motta, 1967). When a fungal colony is growing on a nutrient agar, the rhizomorphs produce branches and the density of branching increases with the concentration of nutrients in the medium up to an optimum value. On such a nutrient medium, every rhizomorph eventually produces a mantle of fringing myce-lium, which further enhances its resemblance to a root with its mantle of root hairs. Fig. 17, after R. Hartig, shows the apical region of a rhizomorph of *A. mellea*.

Much has been learnt about the behaviour of rhizomorphs of *A. mellea* through simple experiments on the nutrient agar plate (Garrett, 1953). When a 4 mm agar disc taken from the growing margin of a colony of *A. mellea* is laid at the centre of a fresh plate of nutrient agar, an outwards growth of unorganized mycelium can be observed within 24 h at $25°$ C; after 7 days, a circle of rhizomorph initials can be seen around the margin of the original inoculum disc. In my own experiments, the young rhizo-morphs, which started growth later than the colony of unorganized mycelium, grew some five times as quickly as the mycelial colony, and hence rapidly outgrew it. From the older region of the rhizo-morphs fringing mycelium grew out, so that a mature rhizo-morphic colony on an agar plate was deeply lobed. My observations were confirmed and amplified by Snider (1959), working with a

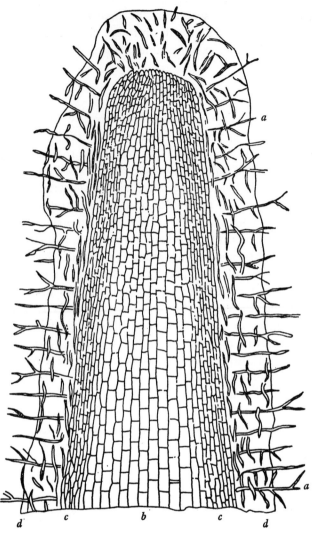

17. Apical region of rhizomorph of *Armillaria mellea*.
(After R. Hartig, 1873.)

number of single-basidiospore isolates of *A. mellea*. Undoubtedly
the most interesting fact to emerge from these experiments was the
wide difference in growth rate between that of hyphae in unorganized mycelium and that of the rhizomorphs; in a subsequent study

by Rishbeth (1968), rhizomorphs of one isolate grew some thirteen times as quickly as the margin of the mycelial colony. A possible explanation for this difference is that the constituent hyphae making up a rhizomorph are unbranched, and so the hyphal apices do not suffer from nutrient competition from the subordinate apices of branches. In a further study on growth of rhizomorphs from small woody inocula through glass tubes of soil, I observed a progressive deceleration with time in the growth rate of rhizomorphs, and obtained evidence that this was due to increasing competition for nutrients from the subordinate apices of rhizomorph branches (Garrett, 1956*b*). But whatever the reason for this very high growth rate of rizomorphs, as compared with that of individual hyphae in unorganized mycelium, it must have one important consequence for the mechanism of ectotrophic infection. If my interpretation of this mechanism, as illustrated in Fig. 12, be accepted, then the faster the rate of epiphytic spread of the rhizomorphs, the more effective will be the ectotrophic infection-habit in overcoming active host resistance.

Another outstanding difference in morphogenesis of mycelial strands and rhizomorphs has also become apparent; whereas strand-building is delayed by the presence in the growth medium of nutrients above a low concentration, the production of rhizomorph initials increases with glucose concentration at least up to the level of 2% (Garrett, 1953). In my own experiments on production of rhizomorphs on nutrient agar, vitamins were provided by addition of thiamin and also by a supplement of malt extract and/or peptone, to guard against any possible vitamin deficiency. Employing a synthetic growth medium with 1 mg thiamin/l as the only vitamin supplement, Weinhold & Garraway (1966) found that rhizomorphs were not produced either in the liquid medium or on the same solidified with agar unless the medium was further supplemented by ethanol, at the rate of 500 ppm. As the medium was provided with 6 g glucose/l as a carbon source, the ethanol supplement was evidently acting as a rhizomorph-inducing factor rather than as a carbon source, even though ethanol by itself at the rate of 4·6 g/l provided a satisfactory carbon source for both mycelial growth and rhizomorph production. Mycelial colonies grew well enough in the absence of ethanol, though growth was usually improved by its addition to the medium. This agrees with our experience at Cambridge; my isolate of *A. mellea* grew well enough in a synthetic liquid medium with thiamin as the only vitamin supplement, but rhizomorphs were virtually absent (Garrett, 1953). My observations with this single isolate were confirmed by Azevedo (1963) with two isolates from England and two from the Azores, respectively. For some isolates, at least, of *A. mellea*, ethanol and other low-molecular-weight alcohols thus seem to act specifically as a rhizomorph-inducing factor, though my own experience suggests that they can be replaced by some unknown factor or factors present in malt extract and in peptone. The results of further experiments by Garraway & Weinhold (1968) showed that for maximum mycelial growth and rhizomorph production, their isolates of *A. mellea* required ethanol for the first 15 days of colony growth; thereafter, lack of ethanol affected neither rhizomorph production nor total growth.

Production of rhizomorphs is not invariably associated with activity of *A. mellea* as a root-infecting pathogen. Their complete absence from areas of root disease in cacao plantations in Ghana was noted by Dade (1927) and their rarity in Malawi by R. Leach (1939). They are abundantly produced in the forests of the Kenya Highlands (Gibson & Goodchild, 1960) but are seldom found in the forests of Rhodesia surveyed by Swift (1968). Where rhizomorphs are thus absent, *A. mellea* spreads from infected to healthy roots by direct contact, and thereafter by mycelial sheets in the bark and over the plane of the cambium; the mechanism of this spread as an alternative to ectotrophic infection by initially epiphytic rhizomorphs clearly calls for further investigation. In an endeavour to elucidate this variability in the production of rhizomorphs by *A. mellea* in different areas, Gibson (1961) studied 140 isolates of *A. mellea* from various parts of the world and found that they varied widely in the regularity, abundance and vigour of rhizomorph production under optimum conditions of pure culture on nutrient agar. A total of 116 isolates was tested for their temperature optima over the three ranges of 22–4° C, 25–7° C and 27–9° C, respectively. Both for mycelial growth and for initiation and growth of rhizomorphs, most isolates had their optima within the range 25–7° C; the frequency distribution of temperature optima did not differ significantly between isolates from temperate and from tropical regions, respectively. These results thus seemed to give little support to the idea that production of rhizomorphs under natural conditions was limited by soil temperature, i.e. that too high a temperature prevented their initiation at low altitudes in the tropics, but not at higher elevations in the tropics nor in temperate climates.

Fortunately, however, this proposition has recently been reinvestigated by Rishbeth (1968), who has thereby brought to light a critical difference between temperature optima for initiation and growth of rhizomorphs on nutrient agar and from woody inocula through soil, respectively. Growth rate both of purely mycelial colonies and of rhizomorphs on 3 % malt extract agar was found to be maximal at a temperature of approx. 25° C; growth rate of rhizomorphs from woody inoculum segments through soil tubes was maximal at approx. 22° C (Fig. 18).

The optimum temperature for growth rate of rhizomorphs from a food base through soil is thus lower, by some 3° C, than the optimum for growth of rhizomorphs on malt-extract agar. I myself had earlier found that growth rate of rhizomorphs from woody inoculum segments through soil tubes is nutrient-dependent. Growth is faster from large inocula than from small ones, and weekly growth increments of rhizomorphs invariably decrease with time from the beginning of growth; I suggested that this latter effect was due partly to a decrease in rate of mobilization of nutrients from the woody food base, and partly to an increasing intensity of competition for nutrients between the main apices of the rhizomorphs and the growing number of their subordinate branch apices (Garrett, 1956b). Rishbeth has used this evidence in support of his suggestion that at the optimum temperature of approx. 22° C, rhizomorph growth rate is in balance with translocation from the food base; at higher temperatures, rate of translocation is insufficient for the higher rate of metabolism in the rhizomorph apices. This suggestion was supported by the effects of transferring soil tubes from a lower to a higher temperature and vice versa. It also explains Rishbeth's observation that the optimum temperature for growth of rhizomorphs on malt-extract agar was 25° C and sometimes higher; under these growth conditions, rhizomorph growth rate would be unlikely to be nutrient-limited, because uptake of nutrients can take place along the whole length of the growing rhizomorph. The optimum temperature for extension of mycelial sheets between bark and wood in cut lengths

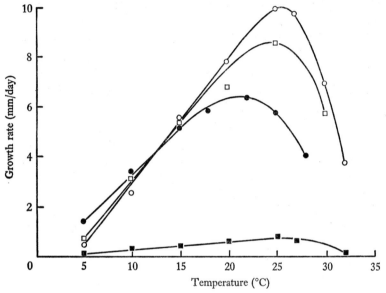

18. The growth rates of mycelium and rhizomorphs of *A. mellea* in relation to temperature. ○, Rhizomorphs in malt agar; ●, rhizomorphs in soil tubes; □, mycelial sheets in wood lengths; ■, mycelium in malt agar. (After J. Rishbeth, 1968.)

of small stems of *Acer pseudoplatanus* was also found to be near 25° C (Fig. 18); under these conditions also, growth rate was unlikely to have been limited by translocation rate, because the mycelium was in direct contact with its substrate for the whole extent of its growth.

In the preparation of his woody inoculum segments, Rishbeth employed a dose of basidiospores as inoculum so that his population of *A. mellea* must have been genetically variable and therefore a representative sample of the population in the area from which he obtained fructifications. Yet in no case did he observe rhizomorph initiation at a temperature above 25° C; from his overall experience in this study, he selected a temperature of 18–20° C as the most reliable for rhizomorph initiation. His total experimental results thus clearly suggest that too high a soil temperature is likely to be one of the factors restricting production of rhizomorphs by *A. mellea* in the lowland tropics.

Rishbeth has, however, disclaimed any suggestion that supra-optimal soil temperature is the only factor limiting rhizomorph production in the lowland tropics, and has drawn attention to the work of Swift (1968) on the existence of a toxic factor in some Rhodesian forest soils. All the isolates of *A. mellea* employed by Swift produced abundant rhizomorphs on nutrient agar, so that an intrinsic inability to produce rhizomorphs could be eliminated as a possible factor in their behaviour. Moreover, although no rhizomorphs were produced from woody inoculum segments in growth tubes filled with a typical Rhodesian forest soil, they were initiated and grew well in the same soil after it had been autoclaved. Soil fumigation with propylene oxide did not remove the rhizomorph-inhibiting factor, so Swift was able to conclude that it was persistent in the soil in the absence of living micro-organisms, at least for a time, though it was not thermostable because it was destroyed by autoclaving the soil. The persistence of this factor was also demonstrated in another way, by Seitz-filtration of a 1:1 soil–water extract and addition of the sterile filtrate to columns of autoclaved soil in the growth tubes; this sterile filtrate caused total inhibition of rhizomorph growth.

Swift also considered another possibility, arising from my observation that rhizomorphs failed to make any appreciable growth into the short column of moist sand with which I covered inoculum segments, standing above soil columns in the growth tubes, in some of my experiments (Garrett, 1956*b*). I had thought that perhaps sand failed to provide some factor present in soil that stimulated outwards growth of rhizomorphs from the inoculum segments; this possibility was later eliminated by Rishbeth's (1968) finding that rhizomorphs grew well enough from the inoculum segments when these were buried in 1 lb glass jars of sand. A satisfactory resolution of this discrepancy has now been provided by Griffin (1969), who has suggested that my observations can be explained by a drying-out of the short sand columns covering the inoculum segments at the top of my growth tubes. In this supposition he has been entirely correct; I can indeed confirm that the short sand columns did dry out fairly quickly, in spite of the fact that the growth tubes were capped by moisture-proof, lacquered cellophane secured with a rubber band. The correct explanation did not occur to me because the drying sand still acted as an effective mulch to prevent drying out of the inoculum segments, and growth of rhizomorphs *downwards* into the soil columns was very satisfactory. But failure of rhizomorph initials to continue growth into the drying sand columns can now be explained by Griffin's discovery, in collaboration with A. M. Smith, that rhizomorphs will continue to grow only for so long as the apices are covered by an unbroken film of water. They have interpreted this requirement in terms of the reduced rate of melanin formation under the lower oxygen concentration around the rhizomorph apex when it is covered by a water film; their explanation is compatible with the high values of the Michaelis constants for the enzymes laccase and tyrosinase. They postulate that the melanin sheath is mycostatic, protecting the mature part of the rhizomorph but inhibiting elongation of the apex if formed precociously.

SOME EFFECTS OF SOIL CONDITIONS ON ECTOTROPHIC SPREAD

Three ectotrophic root-infecting fungi are known to spread more quickly over the root systems of their hosts in alkaline soils than in neutral or acid ones. No explanation seems to have been advanced for this effect on the spread of *Phymatotrichum omnivorum*, but attempts have been made to explain a similar effect on the spread of *Ophiobolus graminis* and *Fomes annosus*, respectively.

During my own investigation of the effect of soil conditions on rate of ectotrophic spread by *Ophiobolus graminis* along the roots of wheat seedlings, I found that rate of spread was increased: (1) by improvement in soil aeration; (2) by increase in pH value of the soil up to approx. pH 8 (Garrett, 1936). In a hypothesis attempting to unify these two sets of observations, I suggested that rate of advance of *O. graminis* was inversely proportional to the concentration of carbon dioxide at the surface of the respiring root; thus at pH 8, 3 % undissociated carbon dioxide is in equilibrium with 97 % as bicarbonate ion, whereas at pH 5·5, 90 % free carbon dioxide is in equilibrium with 10 % as bicarbonate. In further experiments to test this hypothesis, I demonstrated that growth of *O. graminis* along the roots was greatly accelerated by forced aeration of the soil and that under these conditions soil pH value no longer affected the rate of fungal advance (Garrett, 1937). This simple explanation has recently been challenged by the theoretical considerations and experimental results published by Greenwood & Nye (1968). They have pointed out that as diffusion through the gas phase is extremely rapid, oxygen and carbon dioxide partial pressures in any soil gas space that is directly connected with the atmosphere above the soil will be much the same as in air. In the soil–water phase, carbon dioxide is thirty times as soluble as oxygen and so will diffuse thirty times as fast. From this it could be concluded that a deficiency of oxygen at the root surface, due to both root and microbial respiration, is likely to limit the growth rate of *O. graminis*, and of similar ectotrophic root-infecting fungi, *before* the partial pressure of carbon dioxide has increased to a growth-limiting value. Nevertheless, convincing as these arguments by Greenwood & Nye may seem, it is perhaps yet premature to come to any definite conclusion; their arguments challenge not only my own hypothesis but also a substantial volume of other data hitherto accepted as evidence for a growth-limiting effect of carbon dioxide on the activities of various soil micro-organisms.

A different explanation was put forward by Rishbeth (1950) for the more rapid advance of *Fomes annosus* over the bark of pine roots in alkaline as compared with acid soils. He was able to reproduce this effect on 5 cm lengths cut from small pine roots growing in acid and alkaline soils, respectively, washed free of soil, and then incubated, after inoculation with *F. annosus* at one end, in a moist atmosphere. The design of this experiment eliminated any differential effect of carbon dioxide at the root surface, such as I had postulated for *O. graminis*. It must also largely have eliminated any differential effect of active host resistance due to soil type, because as we have noted earlier, Wallis (1961) demonstrated that root-severing destroyed the effective defence-response of roots in either type of soil. The much retarded rate of advance by *F. annosus* over the root bark of roots taken from an acid soil in this experiment was ascribed by Rishbeth to antagonistic effects by root-surface fungi of acid soils, and particularly to those exerted by *Trichoderma viride* Pers. ex Fr. Gibbs (1967) has examined Rishbeth's suggestion of *T. viride* as an important antagonist of *F. annosus* on pine root bark in acid soils in the light of Rifai's (1969) subdivision of *T. viride*, as recognized by Bisby (1939), into nine species. Even within each of Rifai's narrowly defined species of *Trichoderma*, both Gibbs (1967) and Mughogho (1968) have found a wide variation amongst isolates in the degree of antagonistic effects that they exert against other fungi growing in close association with them. Gibbs was unable to find any clear evidence of a significant correlation between frequency of antagonistic isolates of *Trichoderma* and soil acidity. In place of this explanation, he has drawn attention to much published evidence in favour of the generalization that most antibiotics are inactivated much more quickly in neutral and alkaline soils than they are in acid soils.

On the light-textured, sandy soils of the Breckland area of East Anglia particularly studied by Rishbeth (1951*b*), various other soil factors, and particularly organic content, can be closely correlated with soil pH. Thus the acid podsols are covered by a well marked litter layer, and so the moisture retaining and supplying power of these soils in dry seasons is higher than that of the alkaline sandy soils. Rishbeth adduced both observational and experimental evidence for the view that the more favourable moisture régime in the acid podsols enhanced the resistance of pine trees to *F. annosus*, especially in dry seasons. This opinion has been upheld by recent evidence produced by Gibbs (1968), showing that the volume of resin production following injury could be directly correlated with soil moisture availability, and was greater for pine trees on acid podsols than for those on the alkaline sandy soils. In the U.S.A., Towers & Stambaugh (1968) found that experimentally induced moisture stress increased the susceptibility of loblolly pine (*Pinus taeda*) to infection by *F. annosus*.

5 Competitive saprophytic colonization of substrates by root-infecting fungi

This chapter and the two following will be concerned with survival of pathogenic root-infecting fungi in the soil apart from living host plants. As already adumbrated in Ch. 1, there are three possible modes of such survival: (1) by competitive saprophytic colonization of dead organic substrates, most commonly consisting of a corpus of dead plant tissue, and by saprophytic survival on these; (2) by saprophytic survival of the pathogen in dead infected tissues of a host plant, first invaded by the pathogen during its parasitic phase; (3) by dormant survival in the form of resting bodies (spores or sclerotia), which have been produced either in plant tissues colonized saprophytically, as under (1) above, or in tissues infected by the pathogen during its parasitic phase.

Because of the complexity of saprophytic competition for substrates amongst soil fungi, the circumstances under which some root-infecting fungi appear to have a phase of existence as competitive saprophytes in the soil are still incompletely known and imperfectly understood. In the earlier years of root-disease investigation, chance observations presented some valuable clues to a few gifted observers who were alert enough to be thinking about this problem already. In Ch. 2, I have mentioned the evidence for the conclusion that some tropical species of *Rosellinia* have a phase of active saprophytic colonization of leaf litter on the forest or plantation floor; I also described the chain of evidence that permits us to conclude that *Rhizoctonia solani* can exist in many cultivated soils as a free-living competitive saprophyte, for which it is well equipped by its ability to decompose cellulose quite vigorously. Since the beginning of the present century, mycologists from E. J. Butler (1907) onwards have been accustomed to investigate the occurrence of particular groups of fungi in the soil by burying therein 'baits' of particular types of dead plant or animal tissue so selected as to encourage competitive saprophytic colonization by the desired group of fungi, e.g. boiled hemp seed for *Pythium* species, human hair for keratin-decomposing fungi, shrimp or lobster shell for chitin-decomposing fungi, and so on. Sadasivan (1939) baited various soils at the Rothamsted Experimental Station with pieces of fresh, mature wheat straw, and in this way was able to confirm a conclusion from earlier circumstantial evidence that *Fusarium roseum* f. *cerealis* (syn. *F. culmorum*) is a regular primary fungal colonizer of this substrate when it is introduced into cultivated soils.

Isolation of fungi from such baits of dead plant or animal tissue buried in natural and otherwise untreated soil provides the most direct evidence obtainable that such fungi can colonize these particular substrates in soil as competitive saprophytes; for root-infecting fungi, it constitutes evidence for the possibility of an independent saprophytic phase in the soil. This realization provided the basis for a technique for comparing degrees of competitive saprophytic ability amongst root-infecting fungi that was devised and developed by my research associates and myself at the Cambridge Botany School from 1950 onwards; it has been employed by workers elsewhere under the designation of the 'Cambridge Method' and I shall describe it in detail later in this chapter. Like many other experimental methods, this one purports to create a model of the natural situation or situations in the soil, and so invites questions as to its veracity. Two questions in particular need answering. First, do substrates of the type employed experimentally occur at all frequently in the soil? Secondly, when soil artificially inoculated with a particular fungus is employed, does the range of inoculum dosage correspond with the range of population level at which the fungus occurs naturally in cultivated soils? This second question arises from our early discovery that the inoculum potential of a fungus at the surface of a substrate partly determines the share of the substrate that it is able to colonize and exploit.

For these reasons, interpretation of experimental findings on saprophytic behaviour of root-infecting fungi has been hampered by doubts concerning the validity of the experimental models of a kind that has scarcely vexed investigation of other phases of the life-cycle. Thus the parasitic phase, saprophytic mycelial survival in dead, infected host tissues, and survival of dormant propagules all present problems that are much less complicated than that of saprophytic competition, so that no one has really needed to ask whether the experimental models are satisfactory replicas of the natural situation.

From discussions of the parasitic phase in the three preceding chapters, we can conclude that opportunities for purely saprophytic colonization of the root system are probably very restricted by comparison with the volume of root tissue that is infected during the parasitic phase. Infection of living tissues continues into, or may even be initiated during, senescence whether natural or infection-induced; it is finally halted only when tissue-resistance to invasion has declined to a level at which any obligate saprophyte can invade and occupy the remaining volume of root tissue. Living root tissues are reserved for the parasites, whereas dead tissues are open to invasion by all soil fungi; both in time and in space, therefore, opportunities for parasitic invasion by root-infecting fungi greatly exceed their opportunities for saprophytic colonization, and this difference is greatest for the most specialized parasites, because these are the ones least effectively restricted by host resistance during their parasitic phase.

So far as occupation of the root system is concerned, therefore, we can conclude that saprophytic colonization by root-infecting fungi is of minor importance by comparison with infection of living tissues and that for the specialized pathogens it is of negligible importance. But shoot systems of plants eventually reach the soil, and are indeed regularly incorporated therein in the normal course of agriculture; what opportunities does this give for saprophytic colonization by root-infecting fungi as an additional means of multiplication and survival? Hudson (1968) has recently produced a very timely and comprehensive review of investigations conducted, particularly during the last two decades, into colonization of plant shoot systems by air-borne fungi. From this it appears that much of the plant shoot system is colonized by fungi whilst it is still erect; specialized parasites invade the tissues whilst they are still in their prime, weak parasites follow during the phase of senescence, and even saprophytes can enter and colonize before the shoot system reaches the ground. This air-borne fungal colonization will begin earlier and extend further in a moist than in a dry climate, and its extent will increase with the time that the plant shoot system is allowed to remain standing. In a warm, dry climate or season, for instance, the straw of a wheat crop that is ploughed in before the first rains of autumn may consist largely of virgin, undecomposed plant tissue; in a moist climate or season and where ploughing has been delayed, the value of the straw as a substrate for soil fungi will have been correspondingly reduced by prior fungal colonization. It is important to keep in mind this limitation upon the availability of substrates for soil fungi, before we proceed to discuss what factors are likely to determine success in saprophytic competition. For any particular fungus, the share of a potential substrate that it can obtain will be determined: (1) directly by its *competitive saprophytic ability*, which is an intrinsic characteristic of the fungal species; (2) directly by its inoculum potential at the surface of the substrate; (3) inversely by the inoculum potential of its competitors.

COMPETITIVE SAPROPHYTIC ABILITY

I first proposed this term in 1950 and later (1956a) attempted a definition of it as 'the summation of physiological characteristics that make for success in competitive colonization of dead organic substrates'. A definition for a new concept has something in common with a specific binomial for a newly discovered organism; the less one knows about it, the more important it is to be able to identify it. This definition is chiefly revealing about our state of ignorance concerning competitive saprophytic ability at that time. Since then, research has clearly demonstrated that the possession of competitive saprophytic ability is not, as I had originally visualized it, the key to success in competitive colonization of substrates of all kinds; instead, we are beginning to realize that the kinds of competitive saprophytic ability are almost as diverse as the types of substrate open to saprophytic colonization. An obvious parallel is the diversity of athletic prowess that makes for human success in athletic competitions; a champion weight-lifter is unlikely to excell in short-distance sprinting and vice versa. A closer analogy for plant pathologists is to be found in the term 'pathogenicity', which is host-specific, and most of all so in specialized parasites. Just so, competitive saprophytic ability is substrate-specific. In the case of obligately saprophytic fungi, competitive saprophytic ability for the colonization of one or more types of substrate can be equated quite precisely with the general fitness of the fungal species for existence in one or more ecological niches, and the term thus appears to be redundant. But in the case of root-infecting parasites, for which I proposed this term in the first place, it may be conceded a certain usefulness for the answer of certain questions such as the following. Can this fungus exist as a free-living saprophyte in the absence of host plants? If so, how well can it compete with obligate saprophytes? Is the possession of competitive saprophytic ability incompatible with specialization to a parasitic existence? The evidence to be presented later will show that we have made some progress towards an answer to these various questions.

To forestall possible confusion, it is now necessary to point out that the term 'substrate' is widely used in two senses, though the particular sense is usually evident from the context: (1) in a wide sense, to denote a corpus of material, usually of plant tissue, which a fungus has colonized and on which it is subsisting; (2) in a narrow sense, to indicate a particular constituent of plant tissue that is being decomposed by a particular fungus, e.g. a sugar, pectic substance, hemicellulose, cellulose or lignin. This latter sense is similar to that in which the term substrate for an enzyme is employed. When exploiting a corpus of plant tissue as a substrate (in the wide sense), a fungus may be utilizing several constituents of the tissue (i.e. substrates in the narrow sense) simultaneously; the biochemical versatility of the whole organism is wider than that of any one of its enzyme-sequences. 'Saprophytic sugar fungi' have been defined (Burges, 1939) as those able to utilize only sugars and carbon constituents of plant tissues that are simpler than cellulose, i.e. they are incapable of degrading cellulose or lignin. Such a fungus may have a high degree of competitive saprophytic ability for colonization and exploitation of those carbon compounds that it can decompose but will clearly have none for cellulose or lignin, which it cannot degrade even in pure culture. Conversely, a cellulose-decomposing fungus with a high degree of competitive saprophytic ability for this substrate may succeed in securing only a small proportion of the sugars and simpler carbon compounds present in a corpus of fresh plant tissue buried in soil, because its growth rate is lower than that of the fast-growing saprophytic sugar fungi which are usually the pioneer colonizers; they get there first and may consume most of the simpler carbon compounds before slower growing cellulose-decomposers can establish themselves on the substrate.

It is certainly possible to compile a list, as I attempted to do in 1950, of all the physiological and biochemical attributes that could conceivably contribute to competitive saprophytic ability of one kind or another; few successful saprophytes will need, or are likely, to possess all these attributes in high degree, and the particular spectrum of attributes will be determined by the substrate. A short list of the more obvious attributes might be as follows:

(1) rapid germination of fungal propagules and speedy growth of young hyphae when stimulated by soluble nutrients diffusing from a substrate;

(2) appropriate enzyme equipment for decomposition of the more resistant carbon constituents of plant tissues, such as cellulose and lignin;

(3) excretion of fungistatic and bacteriostatic growth products, including antibiotics;

(4) tolerance of fungistatic substances produced by other soil micro-organisms.

The term 'antibiotic' is generally reserved for microbial growth-products active against other micro-organisms at a concentration of 10 ppm or less, and somewhat species-specific in their growth-stasis effects. The term 'fungistatic substance' is usually used with reference to a microbial growth-product that is active only at a significantly higher concentration than the minimum for antibiotic action, and it is non-specific in its effects; examples are carbon dioxide, the bicarbonate ion, and substances generating extremes of pH in the ambient solution.

I will briefly indicate the way in which these various attributes of a fungal species contributing to its competitive saprophytic ability may be distributed amongst various substrate-groups of soil fungi, in the form of a tabulation as shown:

	Speedy germination and growth	Enzymes for cellulose and/or lignin degradation	Production of antibiotics etc.	Tolerance of antibiotics etc.
Primary saprophytic sugar fungi, as pioneer colonizers of fresh plant-tissue substrates	+	—	—	—
Secondary saprophytic sugar fungi, associated with cellulose and lignin decomposers	—	—	+	+
Root-surface and rhozisphere fungi	—	—	+	+
Cellulose and lignin decomposers	—	+	+	+

Some of the evidence for the conclusions summarized in this tabulation will be presented later in this chapter, but a brief commentary will be helpful at this stage. The primary saprophytic sugar fungi are, by definition, pioneer colonizers, and they occupy their substrate ahead of competitors so that neither production of nor tolerance to antibiotic substances is likely to give them any significant competitive advantage. The other three substrate-groups live in an environment in which they are exposed to the full intensity of microbial competition, so we can expect that both production of and tolerance to antibiotic substances will confer an additional degree of competitive saprophytic ability.

Secondary saprophytic sugar fungi are designated as those that live as commensals with cellulose- and lignin-decomposing fungi, i.e. on the cellobiose and glucose released by hydrolysis of cellulose, and on soluble products of lignin breakdown. Arising from my particular interest in the utilization of cellulose during saprophytic survival of cereal foot-rot fungi in infected wheat straw (Ch. 6),

I postulated the existence of this commensal group of secondary saprophytic sugar fungi (Garrett, 1963, p. 100). The occurrence of this type of commensalism has been most elegantly demonstrated by Tribe (1966). He was able to grow several species of *Pythium* (sugar fungi) on cellulose film in association with various species of cellulose-decomposing fungi; the amount of growth made by a species of *Pythium* in presence/absence of an associated cellulose decomposer was estimated quantitatively by oospore counts.

Commensalism of this type has also been demonstrated by Chang (1967) for another non-cellulolytic fungus, *Humicola lanuginosa*, when grown in association with the strongly cellulo-lytic *Chaetomium thermophile* on filter-paper cellulose. When *C. thermophile* was grown by itself, soluble products of cellulolysis accumulated in the filter-paper pads, and the presence of reducing sugars was demonstrated; when grown in association with *H. lanu-ginosa*, weight of solutes was reduced, and only a trace of reducing sugars could be detected. This and other evidence on the ecological niche occupied by secondary saprophytic sugar fungi has been recently discussed by Hudson (1968).

INOCULUM POTENTIAL

I have already defined and discussed inoculum potential as a determinant of infection in Ch. 1 (p. 9). With a slight change in wording, this definition can be made to apply equally well to competitive saprophytic colonization: *inoculum potential is defined as the energy of growth of a fungus available for colonization of a substrate at the surface of the substrate to be colonized* (Garrett, 1956a, p. 41). It is perfectly true that the population level of a particular fungus in natural soil, or its inoculum dosage in an experiment, is one of the most important factors determining the share of a substrate that will be obtained by that fungal species in competition with others. But it is not the only factor, because the success of a population of fungal propagules in competitive saprophytic colonization will also be affected by other factors, both endogenous and exogenous. The growth vigour of the propa-gules will be determined partly by their age and nutrient status (endogenous factors) and partly by nutrients diffusing from the surface of the substrate and by environmental conditions (exogenous factors). This whole complex of factors affecting the energy of growth of a fungus at the surface of a substrate, and similarly that of its competitors, is summarized by the concept of inoculum potential, and so this concept is just as apt for research on saprophytic colonization as for research on the process of infection. In the broad sense of the term 'competition', fungi compete for a share of substrate not only with other fungi, but also with bacteria and actinomycetes, and at least with members of the soil microfauna (Ch. I, p. I) amongst soil animals. Nevertheless, it is axiomatic in the science of ecology that the keenest competition occurs between organisms occupying the same or closely similar ecological niches; for this reason, no doubt, all the investigations to be discussed in this chapter have been restricted to competition between fungi, and therefore my discussion of microbial competition will be restricted to saprophytic competition between fungi.

Although I am using the concept of inoculum potential in order to elucidate the mechanism of saprophytic competition, we have to distinguish carefully between its role in this situation, and that as a determinant of root infection. In the process of infection, inoculum potential of the fungus is the force that opposes resistance of the host; only if the inoculum potential of the fungus is sufficient to overcome resistance of the host does a successful, progressive infection ensue. A substrate, on the other hand, is inert and offers no active resistance to colonization; the resistance to be overcome is constituted by the competition of other organisms. For this reason, the propagule-population (per unit area of host or substrate surface) component of inoculum potential appears to be relatively more important in competitive saprophytic colonization than it is in host infection. Perhaps I can make this more clear by an analogy with human behaviour in a situation all of us must have experienced. In some countries at all times, and in all at some times, there is disorderly behaviour amongst a crowd of intending passengers trying to get into a public-transport vehicle. The door is open, so the vehicle offers no 'resistance' to invasion; for any one intending passenger, the only resistance comes from the competition of others. But success does not necessarily go only to the strongest; a person who happens to be immediately opposite the open door when the vehicle comes to a halt may be able to board it virtually without opposition, provided always that the crowd accepts a certain minimum level of civilized behaviour. Thus an 'advantage of position' may more than compensate for physical weakness; just so, a high propagule-population of a particular fungal species at the surface of a substrate gives the species an advantage of position which can more than compensate for a low degree of competitive saprophytic ability. If thus favoured by a high population level, almost any fungus can obtain some share of a waiting substrate, as we shall see shortly.

In conclusion, we can epitomize the mechanism of saprophytic competition between fungi in the following statement: *the share of a substrate obtained by any particular fungal species will be determined partly by its intrinsic competitive saprophytic ability and partly by the balance between its inoculum potential and that of competing species.* Environmental conditions in the soil, as we shall see, can strongly affect the outcome of saprophytic competition; they do this by altering the balance between inoculum potentials of the competing fungi. Saprophytic competition between fungi presents some of the most difficult problems in the whole field of root-disease investigation; for this reason, I have thought it best to outline enough of what we have learnt about saprophytic competition as will be necessary to help comprehension of the experimental evidence, now to follow.

THE CAMBRIDGE METHOD

The Cambridge method for investigating competitive saprophytic colonization of substrates by root-infecting fungi, which I adumbrated in its original form in 1950, has been developed and employed by the following of my former research associates: F. C. Butler (1953 *a*, *b*), R. L. Lucas (1955), A. S. Rao (1959), R. C. F. Macer (1961 *a*) and R. L. Wastie (1961). Pure cultures of a fungus to be tested were grown upon a mixture of quartz sand

+3% maizemeal; when the culture was fully grown, it was easily miscible with soil in any desired proportion. In preparation for an experiment, the maizemeal–sand culture of the fungus was progressively diluted with an unsterilized soil of medium texture to give a dilution series; a series that we commonly employed contained the following percentages of pure-culture inoculum: 100 (pure culture alone), 98, 90, 50, 10, 2 and 0 (unsterilized soil alone). Several glass jars were filled with each of these mixtures; in each jar were then buried fifty standard *substrate units* of dead plant tissue. The substrate units were incubated in the inoculum–soil mixtures for about 4 weeks, so as to give the inoculant fungus ample time to colonize as many of them, in competition with other soil fungi, as it was capable of doing. At the end of this period, the number of substrate units colonized by the inoculant fungus in each of the inoculum–soil mixtures was determined.

For investigation by the Cambridge method, we selected four (subsequently five) cereal foot-rot fungi, because at that time most of my previous research experience had been with this group of pathogens. These fungi were *Fusarium roseum* f. *cerealis* (syn. *F. culmorum*), *Ophiobolus graminis*, *Cochliobolus sativus* (Ito & Kuribay.) Drechsl. ex Dastur (stat. conid. = *Helminthosporium sativum* Pamm., King & Bakke), and *Curvularia ramosa* (Bain.) Boedijn. At that time, there was already good evidence for supposing that *F. roseum* was a vigorous competitive saprophyte and that *O. graminis* was a poor one; these two fungi thus provided reasonably reliable points of reference by which to judge the value of the method. The choice of this group of fungi had the further practical advantage that the host-plant tissue, which seemed the most appropriate to use in these tests, could be wheat straw, which was easily obtained and cut up into standard substrate units; short lengths (approx. 4 cm) were so cut from the long straw as to have a node at the lower end, and this served to hold the straw pieces together during subsequent handling. Before burial in the inoculum–soil mixtures, the straws were soaked in water or various nutrient solutions, and then autoclaved.

Various methods were tested for assessing the percentage of substrate units colonized by the inoculant fungus after the 4-week incubation period in the series of inoculum–soil mixtures. Washing, followed by surface-sterilization and plating out on nutrient agar, was found to be a reliable method only for fungi that have a high degree of competitive saprophytic ability for colonization of nutrient agar, like *F. roseum*; fungi with a low degree of competitive saprophytic ability for this substrate are suppressed by competitors, as will be described in more detail later in this chapter. A perfectly reliable, though laborious, method was the wheat seedling test I had originally devised for determining longevity of saprophytic survival by *O. graminis* in similar pieces of infected or colonized wheat straw (Garrett, 1938). A wheat seed is placed inside the lumen of each piece of wheat straw to be tested for presence of viable *O. graminis*; a number of such inseminated wheat straws are then planted together in a box of sand, which provides a rooting medium optimum for infection by *O. graminis*. Viable *O. graminis* is subsequently demonstrated by the occurrence of runner hyphae and lesions on the roots of the wheat seedlings after approx. 3 weeks; the method is suitable also for detection of other cereal root-rot fungi. A second reliable method was the 'sand plate' devised by F. C. Butler (1953a) by which wheat straws were recovered from the inoculum–soil mixtures, washed and plated out on a layer of moist sand in Petri dishes; Macer (1961a) subsequently found that surface-sterilization after washing the straws was not only unnecessary but was actually harmful in partially suppressing the inoculant fungus, and so was best omitted. The sand plate method is merely a variant of the 'moist chamber' method long employed by mycologists to encourage fungi to sporulate and thus reveal their presence within infected or colonized plant tissue. Sporulation on the surface of a colonized substrate within a moist chamber does not involve any appreciable degree of competition between fungi *already* established on the substrate, and so the method does not have the disadvantage of strong selectivity, as does plating out on nutrient agar. It is most satisfactory for fungi like *C. sativus* and *C. ramosa* that produce large, dark coloured spores that are easily visible under the stereoscopic microscope.

The results obtained by the Cambridge method for these four

fungi can be illustrated by a selection from the data of F. C. Butler (1953 a). Butler tried all three methods described above for assessing percentage of straws colonized by his inoculant fungi; I have selected for each fungus the data from an experiment in which assessment of colonization was made by the wheat seedling test. As I have pointed out elsewhere (Garrett, 1967), the dilution end-point is likely to be lower for infection of wheat seedling roots than for sporulation in a moist chamber, so it is justifiable to compare only results obtained by the same technique, as I have done in Table 4.

Table 4. *Saprophytic colonization of wheat straw by four cereal foot-rot fungi*

(From F. C. Butler, 1953 a)

	Percentage straws colonized						
Percentage inoculum	100	98	90	50	10	2	0
Fusarium roseum	90	87	80	67	65	55	22*
Curvularia ramosa	100	96	96	96	85	88	0†
Cochliobolus sativus	100	33	46	10	8	2	0
Ophiobolus graminis	98	46	8	10	2	0	0

* Colonized by *F. roseum* propagules occurring in the soil.
† *C. ramosa* does not occur naturally in English soils.

As we had expected to find, the percentage of straws colonized by each inoculant fungus declined progressively as the inoculum was diluted with unsterilized soil; this not only diluted the inoculum but provided an increasing degree of competition from soil fungi and other micro-organisms with the increasing proportion of unsterilized soil. But as we had hoped to find, the *rate* of this decline with dilution of the inoculum provided an estimate of the degree of competitive saprophytic ability in the four inoculant fungi under comparison. Decline in percentage straws colonized by *F. roseum* and *C. ramosa*, respectively, was quite shallow down the dilution series; in contrast, decline by *O. graminis* and *C. sativus*, respectively, was comparatively steep. We thus felt justified in concluding that the first group possessed a high degree of competitive saprophytic ability for colonization of wheat straw, whereas the second group possessed a much lower degree, at least under the conditions of these tests. These results for *F. roseum* and *O. graminis*, respectively, agreed with expectation from earlier evidence on their competitive saprophytic abilities.

In a second paper, Butler (1953*b*) gave the results of various experiments designed to elucidate the difference in competitive saprophytic ability between *C. sativus* and *C. ramosa*. These two fungi had a growth rate over the surface of nutrient agar of a comparable order. Tests for antibiotic production by the stronger saprophyte, *C. ramosa*, were negative, whereas both isolates of *C. sativus* produced antifungal culture filtrates; this difference was thus the reverse of that required to explain the difference in their respective competitive saprophytic abilities. But a possible explanation was furnished by the finding that *C. sativus* was much more sensitive to antifungal antibiotics, of which ten were tested, than was *C. ramosa; C. sativus* was also more sensitive to antagonism by some common soil bacteria in culture.

A fortunate discovery by Lucas (1955) led to an important new insight into the nature of saprophytic competition during the invasion of wheat straw tissue. Quite by chance, he discovered that, in straws from all the inoculum–soil mixtures he tested, *O. graminis* was more abundant in the outer than in the inner tissues of the straw. He interpreted this observation by supposing that the struggle for colonization of the straw between *O. graminis* and other soil micro-organisms in the soil component of the mixture was not settled once and for all in the outer layer of straw tissue; competition was postulated as continuing just as intensely whilst the organisms progressed inwards with the result that *O. graminis*, as a fungus of low competitive saprophytic ability, got left behind. If this supposition were correct, then we should predict that a vigorous competitive saprophyte like *F. roseum* should not get left behind in this way but should be as abundant within the inner as in the outer tissues. This point was verified by Macer (1961*a*) and so these observations have confirmed the difference in saprophytic ability between these two fungi in another way. This finding led Macer to determine respective rates of straw tissue penetration in pure culture by these four fungi, together with a fifth, *Cercosporella herpotrichoides* Fron, which causes the eyespot lodging disease of cereals. The difference in straw penetration rates by *O. graminis* and *F. roseum*, respectively, turned out to be greater than that between their respective surface growth rates over nutrient agar.

Sufficient evidence has now accumulated for an attempt to interpret the marked difference in competitive saprophytic ability for straw tissue colonization between these two pairs of fungi: the vigorous saprophytes, *F. roseum* and *C. ramosa*, and the weak saprophytes, *O. graminis* and *C. sativus*. The data assembled for this purpose in Table 5 are: (*a*) the approximate percentage of unsterilized soil in the inoculum–soil mixture that permits 50% of the straws to be colonized by each inoculant fungus (Table 4); this gives a single-figure estimate (the ED50) for competitive saprophytic colonization; (*b*) superficial growth rates of the four fungi over nutrient agar, taken from Macer (1961*a*); (*c*) figures for straw penetration rate in arbitrary units, taken from Macer (1961*a*); (*d*) a *reduction factor* for each fungus, expressing the proportion by which growth is likely to be reduced by antagonistic effects from competing micro-organisms. These values of the reduction factor for each of the four fungi have been taken from results obtained with a cellophane-covered agar plate test devised by Wastie (1961), which will be described in the next section of this chapter.

It can be seen in Table 5 that there is no correlation between saprophytic success (*a*) and growth rate over nutrient agar (*b*). Straw penetration rate (*c*) is virtually the same for three of these fungi, but *O. graminis* is handicapped by a straw penetration rate only about 0·6 that of the others. But if values for straw penetration rate (*c*) are adjusted by multiplying by the reduction factor (*d*), to allow for the respective proportions by which growth rate through the straw tissue is likely to be reduced by fungistatic effects from other soil micro-organisms, then we get the product (*cd*) shown in the fifth column of Table 5. Mean values of *cd* for the two groups of vigorous and weak saprophytes, respectively, are correlated well

Table 5. *Distribution of characteristics possibly determining competitive saprophytic ability for colonization of wheat straw tissue amongst four cereal foot-rot fungi*

	(a) Percent. unster. soil permitting colonization of 50% straws by inoc. fungus at lab. temp. (16–22° C)	(b) Growth rate over agar at 22·5° C (mm/24 h)	(c) Straw penetration rate at 22·5° C (arbitrary units)	(d) Reduction factor due to competitors	(cd)
Fusarium roseum	approx. 98	10·8	10·0	0·87	8·7
Curvularia ramosa	> 98	4·8	10·5	0·51	5·4
Cochliobolus sativus	approx. 10	3·8	10·3	0·16	1·6
Ophiobolus graminis	approx. 2	8·0	6·4	0·24	1·5

enough with mean values for saprophytic performance. We can conclude that the poor performance of *C. sativus* under these experimental conditions can be ascribed chiefly to its low tolerance of fungistatic growth products produced by soil micro-organisms: this affords independent confirmation, from Wastie's values for the reduction factor, of Butler's (1953*b*) original conclusion. The poorest saprophyte, *O. graminis*, is handicapped both by low tolerance of fungistats and by a straw penetration rate lower than that of any of the other three fungi.

These experiments at Cambridge were carried out under standardized conditions, so as to be comparable with one another, and the jars of inoculum–soil mixtures were incubated at laboratory temperature (range 16–22° C). But in my discussion of inoculum potential earlier in this chapter, I pointed out that the share of substrate obtained by any particular fungal species would be determined by the balance between its inoculum potential and that of its competitors; we should further expect that this balance will be altered by a sufficient change in environmental conditions, which are one of the determinants of inoculum potential. This expectation has been confirmed by the work of Burgess & Griffin

(1967) at the University of Sydney, employing the Cambridge method with two of our fungi, *Cochliobolus sativus* and *Fusarium roseum* f. *cerealis*, and two others, *Gibberella zeae* and *Cochliobolus spicifer* Nelson. Their experiments were designed to investigate the effect of temperature upon competitive saprophytic colonization by these four cereal foot-rot fungi; the temperatures they employed were 10, 20 and 30° C, together with two programmed régimes of fluctuating temperature, 10 ± 5° C and 30 ± 8° C. All four fungi colonized the maximum percentage of wheat straws at the lowest temperature, 10° C. Their increased success at 10° C cannot be ascribed to a direct effect of temperature on their own inoculum potential, because for all of them the optimum for growth in pure culture was 25° C or above; instead, as Burgess & Griffin have suggested, we must conclude that a soil temperature of 10° C depressed the overall inoculum potential of competing saprophytes more than it depressed that of any of the four fungi under test. This situation thus closely parallels the effect of temperature upon pre-emergence killing of seedlings, elucidated by L. D. Leach (1947) and discussed in Ch. 2; although the optimum temperature for growth of the fungal pathogens is around 25° C, percentage pre-emergence killing in some host–pathogen combinations may be greatest at a much lower temperature, because the decrease in temperature retards velocity of seedling emergence more than it retards fungal activity.

In addition to reporting this important and hitherto undescribed effect of temperature upon competitive saprophytic colonization, Burgess & Griffin have also pointed out that at 20° C, approximating to the mean laboratory temperature at which Butler's (1953a) experiments were incubated, they obtained higher percentages of saprophytic colonization by *Cochliobolus sativus* than Butler did. They suggest that this can be explained by supposing that the intensity of microbial competition was lower in their soil than in that employed by Butler. In a further paper, Burgess & Griffin (1968) reported natural colonization by *C. sativus* of wheat straws buried in two uninoculated soils and incubated at 10° C, although the spore population of *C. sativus* in both soils was too low to cause infection of wheat seedlings. They concluded that, at this temperature, *C. sativus* might be able to increase its population by saprophytic activity, and thereby attain an inoculum potential high enough for infection of wheat seedlings; their suggestion that wheat stubble should be ploughed into the soil in summer, whilst soil temperatures are still high enough to suppress saprophytic activity of *C. sativus*, seems a logical application of their findings to farm practice. A similar effect of temperature upon saprophytic activity of *Ophiobolus graminis* has been found by Gerlagh (1968), also employing the Cambridge method; percentage saprophytic colonization of wheat straws was found to increase down the temperature series 23, 17 and 10° C.

Whereas the results of Burgess & Griffin and of Gerlagh show that our original studies at Cambridge, made over a temperature range of 16–22° C, underestimated the competitive saprophytic activity of cereal foot-rot fungi under a lower temperature régime in the field, the work of Cook & Bruehl (1968) in the Pacific Northwest region of the U.S.A. suggests that another characteristic of the Cambridge method has probably overestimated it; these two opposite effects may to some extent cancel each other out. Selecting clean, bright unsterilized wheat straw for burial in containers of soil held at optimum moisture content and at 20° C, Cook & Bruehl found that 94–8% of straws were colonized by *Fusarium roseum* f. *cerealis* from propagules occurring naturally in the soil, which was not inoculated; determination of *F. roseum* was made by plating out straws on nutrient agar, a method quite appropriate for this particular fungus. Their results from this laboratory test thus repeated ours in Cambridge. But when they made a survey of the fungal population in old straw pieces recovered 2–12 months after ploughing or cultivating into soil in the Pacific Northwest region of the U.S.A., they found that the pattern of colonization of the straw pieces by *F. roseum* closely resembled the pattern of *infection* of the wheat plant by this fungus around harvest time, i.e. *F. roseum* was isolated most frequently from the foot or collar region of the straw and frequency of isolation declined with increasing height above soil level. Cook & Bruehl have therefore concluded that, in their wheat-growing area at least, multiplication of *F. roseum* in its parasitic phase much exceeds its increase in population by competitive saprophytic colonization of wheat straw; this conclusion accords with the more general view I expressed in the preamble to this chapter. These authors have discussed possible reasons for this apparent discrepancy between the results from the laboratory test and the field survey: (1) use of clean, bright straw for the laboratory test, whereas much of the straw ploughed into the soil after harvest may be 'weathered', i.e. precolonized by air-borne fungi; (2) plating of straws within a month of burial in the laboratory test but sometimes longer after ploughing in the field survey; thus Walker (1941) obtained a decreasing percentage isolation of *F. roseum* with time from first burial of straws in soil over the period 4–20 weeks tested; (3) employment of optimum soil moisture content in the laboratory containers, as compared with a drier surface soil after harvest in the field. In a survey of eighty randomly selected fields of wheat, Cook & Bruehl were unable to detect *F. roseum* by plating on a selective agar in fifty-eight fields, and in another thirteen fields the population was < 100 propagules/g soil; according to Cook (1968) these propagules are mainly chlamydospores. They argue that if *F. roseum* were a regular saprophytic colonizer of wheat straw in their area, then a higher and more consistent population level of this fungus in the soil would be found. From the survey made by Snyder & Nash (1968) of populations of *Fusarium* species in the soil of various fields at the Rothamsted Experimental Station, there is no doubt that frequent cropping with cereals does favour the multiplication of *F. roseum* f. *cerealis*. Thus populations of 2–3×10^3 propagules/g soil were characteristic of Broadbalk Field cropped continuously with wheat since 1843, whereas from the soil of Barnfield, under root crops alone since 1876, *F. roseum* was rarely isolated. But what proportion of the population on Broadbalk has come from saprophytic colonization still remains a matter for conjecture.

Thus far we have endeavoured to interpret the available evidence on competitive saprophytic ability for colonization of a wheat straw substrate in terms of speed of propagule germination and fungal growth, production of fungistatic substances and tolerance of those produced by other soil micro-organisms. It is also possible that the rate at which a fungus can decompose cellulose and other more resistant components of wheat straw tissue may be a component of competitive saprophytic ability for this substrate, though this possibility has not been explored; the connection between cellulolysis rate of various cereal foot-rot fungi and their saprophytic survival in infected or colonized wheat straw tissue will be considered in Ch. 6. Yet another possible component of competitive saprophytic ability has been suggested by Byther (1965) and Lindsey (1965) at the University of Colorado. Thus Byther found that *Fusarium roseum* was superior to *F. solani* in its utilization of inorganic nitrogen, and especially of nitrate nitrogen, in pure culture under sub-optimal conditions for growth, such as a low level of carbon nutrient, a high pH value of the growth medium and a low ambient temperature. Lindsey employed this information to interpret his data on saprophytic competition between *F. roseum* and *F. solani* for colonization of autoclaved soil and then for establishment in soil microbiological sampling tubes (Mueller & Durrell, 1957). *F. roseum* was the stronger saprophyte of the two in this experimental situation, but the degree of its dominance over *F. solani* was enhanced by shortage of available nitrogen, and reduced when nitrogen supply was not limiting growth. These results thus suggest that the speed and efficiency with which a fungus can absorb nutrients from a limited supply may be a factor contributing to its competitive saprophytic ability.

COLONIZATION OF THE NUTRIENT AGAR PLATE

This subject has been of interest to two groups in particular amongst mycologists. First, mycologists studying the fungal flora of the soil have employed various types of nutrient agar as a substrate for the isolation of soil fungi. As I have described in detail elsewhere (Garrett, 1963), techniques fall into two groups. In the first group, a soil suspension at a suitable dilution is incorporated with the melted agar to give a soil dilution plate (Hornby, 1969), or finely sieved soil is mixed with melted agar in the Petri dish to give a 'soil plate' (Warcup, 1950). In the second group, nutrient agar is exposed to colonization by soil fungi *in situ* in the soil, as in Chesters's (1940) soil immersion tube and its modifications, the screened immersion plate of Thornton (1952) and the soil microbiological sampling tube of Mueller & Durrell (1957). Secondly, all root-disease pathologists studying root-infecting fungi have frequent occasion to attempt isolation of their primary parasite, either for checking a disease diagnosis or for procuring a fresh isolate of the fungus for study. This isolation is often difficult of accomplishment without special precautions; specialized parasites more often than not have a low competitive saprophytic ability for colonization of the agar plate, and so they are suppressed by unspecialized parasites and saprophytes that have followed them into the root tissues (Sadasivan, 1939). The special precautions that have to be taken for success include selection of recently invaded tissue and its surface-sterilization before plating out, both of which reduce the inoculum potential of competing fungi, and employment of a selective medium and/or temperature of incubation to favour the primary parasite over its fungal competitors. Competing bacteria are more easily suppressed, formerly by acidifying the agar and more recently by incorporation of anti-bacterial antibiotics with the plating medium.

It therefore occurred to us to modify the Cambridge method so as to make a more systematic study of competitive saprophytic colonization of the nutrient agar plate than had hitherto been attempted, and this was done in turn by Rao (1959) and Wastie (1961), employing a range of root-infecting fungi as subjects. In this agar-plate variant of the Cambridge method, the inoculum–soil mixtures were made up as before and sufficient aliquots of each to provide replicates were spread out as a layer 1·0–1·5 mm deep over the bottom of a Petri dish and then impregnated by and covered with water agar, first melted and then cooled down to approx. 40° C. After the agar had solidified, discs of 4 mm diam. were cut out with a cork borer and set out (four to a plate), on five replicate plates of modified Czapek-Dox+0·05% yeast-extract agar. The number of pure colonies of the inoculant fungus arising from a total of twenty discs for each inoculum–soil mixture was then recorded. Partial colonies, i.e. a sector of the inoculant fungus in a composite colony with one or more other fungi, were recorded as fractions, the sum of which was then added to the number of pure colonies established by the inoculant fungus.

An improvement to the original Cambridge method in standardization of the pure-culture inoculum was introduced by Rao and employed by Wastie in all his final assessments of saprophytic performance on the agar plate by the eleven root-infecting fungi that he tested. By the original method, pure cultures of all inoculant fungi on sand $+3\%$ maizemeal were incubated at $25°$ C for about 4 weeks. We had realized that this standard period of about 4 weeks might not coincide with 'peak vigour' of all the fungi grown on the medium, and this period of incubation was fixed thus so as to be long enough for the slower-growing fungi. For all his eleven fungi, therefore, Wastie performed separate colonization experiments with pure cultures ranging in age (at $25°$ C) from 5 days by steps of 5 to at least 30 days. For his final results with each fungus, he then selected that age of the inoculum that had given maximum colonization of the agar plates by the inoculant fungus. In this way, it was possible to compare saprophytic performance by each fungus at its peak inoculum vigour, and so to eliminate one source of experimental error.

Wastie confined his experimental consideration of competitive saprophytic ability to the two characteristics of a fungus that then seemed most likely to determine its success on the agar plate: (1) growth rate in pure culture on the virgin nutrient agar; (2) tolerance of fungistatic microbial growth-products as expressed by the ratio 'growth rate on staled agar/growth rate on virgin agar'. For the purpose of determining these two values for each fungus, Wastie successfully developed an ancillary technique, the cellophane-agar-plate test. First of all, he inoculated an agar plate with soil at four equidistant loci, and then covered the whole plate with a thin sheet of wet cellophane. He then inoculated the surface of the cellophane, immediately above each site of soil inoculum, with an agar inoculum disc (4 mm diam.) of the fungus to be tested. In four series of replicate plates, the fungus was inoculated onto the cellophane at 0, 12, 24 and 48 h after the original inoculation of the plate with soil. By comparing rate of growth of the inoculant fungus over the cellophane on plates that had received no soil inoculum with those over plates receiving soil 0, 12, 24, and 48 h previously, Wastie was able to measure the depression in growth rate due to growth of soil fungi from soil inocula under the cellophane. This depression was ascribed to production of fungistatic growth-products by these fungi, accumulating in the agar and diffusing through the cellophane. In Wastie's Table 2, he has listed for fourteen root-infecting fungi the values of the ratio 'growth rate on 48 h staled agar/growth rate on virgin agar'. Under the name of Wastie's 'reduction factor', I have already made use of this ratio in Table 5; as Wastie's values for all fourteen fungi are of general interest, I have reproduced them here in Table 6.

Table 6. *Tolerance of fourteen root-infecting fungi towards fungistatic microbial growth-products, expressed by Wastie's 'reduction factor'*

Rhizoctonia solani	0·94	*Macrophomina phaseoli*	0·33
Fusarium roseum f. *cerealis*	0·87	*Pythium mamillatum*	0·25
F. oxysporum f. *vasinfectum*	0·80	*Ophiobolus graminis*	0·24
F. oxysporum f. *cubense*	0·66	*Cochliobolus sativus*	0·16
F. avenaceum	0·53	*Verticillium dahliae*	0·11
Curvularia ramosa	0·51	*Helicobasidium purpureum*	0·07
Fusarium caeruleum	0·35	*Fomes annosus*	0·06

Wastie obtained a complete set of the required data for eleven out of the fourteen fungi listed in Table 6. This comprised: (1) a colonization rating for each fungus at peak vigour of its inoculum; the rating was the mean percentage colonization for the whole range of inoculum–soil mixtures, which contained 98, 90, 75, 50 and 10% inoculum, respectively; (2) growth rate of each fungus in pure culture on cellophane overlying virgin agar, and growth rates across the cellophane when the underlying agar had been inoculated with soil 0, 12, 24 and 48 h previously. Wastie set out his final conclusions in the form of four correlation diagrams; I have given a summary of them in Table 7, more concisely and with a test of statistical significance, by means of correlation coefficients.

We can see in Table 7 that values of r, the correlation coefficient, fail to show any significant correlation between the colonization

Table 7. *Success of eleven root-infecting fungi in competitive sapro-phytic colonization of virgin nutrient agar from a series of inoculum-soil mixtures*

(Calculated from data of Wastie, 1961)

Between colonization rating (arcsin-transformed) and the following:	Value of r	Value of p	Signifi-cance of r
Tolerance of fungistatic growth-products (= reduction factor)	0·2547	> 0·1	N.S.
Growth rate on 48-h staled agar	0·5000	0·1	N.S.
Growth rate on virgin agar	0·6631	0·02	At 2 % level

rating and either growth rate of the various fungi on 48 h staled agar or their tolerance of fungistatic growth-products. But the value of r for the correlation between saprophytic success and fungal growth rate on virgin agar is significant at the 2% level. This suggests that the outcome of the struggle between the inoculant fungus and other soil fungi in the inoculum–soil agar disc is decided before any fungistatic concentration of fungal growth-products has had time to accumulate in the agar; the issue may even be decided within the first 24 h after plating, because outwards hyphal growth from the inoculum–soil discs can usually be seen by the end of this period, at least under the microscope. If this assumption is correct, then Wastie would have obtained a still better correlation between saprophytic success and fungal growth rate if he had measured growth of his fungi in pure culture after the first 24 h. For nearly all the fungi that he tested, the first 24 h would have come within the lag phase of growth, i.e. the phase before the steady-state growth rate is attained. The lag phase is made up of the period before the fungus makes any microscopically visible growth together with the ensuing period over which growth rate is gradually increasing to the steady-state phase. For all his fungi except two very fast and one very slow grower, Wastie estimated growth rate by measuring colony diameters after 59 h. This value does not tell us enough about growth over the first 24 h, because the shape of the lag-phase part of the growth curve may vary widely between a number of fungi that have all attained the same colony diameter after 59 h.

We can thus conclude from Wastie's data that for a primary saprophytic sugar fungus acting as a pioneer colonizer of a sugar-based nutrient agar, or of some kinds of dead but fresh plant tissue, the most important component of its competitive saprophytic ability consists in rapid germination and early growth of its propagules when activated by a nutrient stimulus from the substrate. This constitutes part of the evidence for the schema on competitive saprophytic ability of various substrate-groups of soil fungi set out earlier in this chapter (p. 116). This conclusion will not come as a surprise to any mycologist who has studied the soil fungus flora by one of the cultural isolation methods to which I have referred earlier in this section. In recent years, fungistatic substances like rose bengal have been added to nutrient agars employed for isolation of soil fungi, in order to slow down fast-growing fungi so that others get a better chance of developing colonies; this function of rose bengal is additional to its valuable bacteriostatic effect (Smith & Dawson, 1944). Wastie's conclusion is also supported by the findings of Lindsey (1965), who inoculated pairs of fungi simultaneously into autoclaved soil, which was later assayed by the soil microbiological sampling tubes of Mueller & Durrell (1957); the faster-growing fungus of each pair was the one predominantly isolated by this procedure.

Fungal growth rate is also likely to be the paramount determinant in intraspecific fungal competition on the nutrient agar plate or slant tube. In a mass isolation of a fungal pathogen from infected tissue containing more than one wild-type strain of the fungus, the fastest-growing one is likely eventually to suppress the others in prolonged culture over a succession of transfers. This is well illustrated by Stover's (1950) study of the brown and gray wild types of *Thielaviopsis basicola*, causing black root-rot of tobacco. On potato-dextrose agar, the growth rate of the brown wild type was greater by approximately one-third than that of the gray, which tended to be dominated and eventually suppressed by the brown. As I pointed out in Ch. 2, citing Martinson (1963) on *Rhizoctonia solani*, virulence in unspecialized pathogens tends to be associated with a high mycelial growth rate; this association is found also in *T. basicola*, inasmuch as the faster-growing brown wild type is a more virulent pathogen than the gray. Prolonged culture on nutrient agar, with a succession of transfers to fresh media, does not usually result in loss of pathogenicity by stock cultures of unspecialized fungal pathogens; the procedure is selective for the fastest-growing strains in a mixture and therefore also for the most virulent ones. Unfortunately this mechanism operates in the reverse direction for the more highly specialized parasites, in which growth rate tends to be *inversely* correlated with the degree of parasitic specialization (Ch. 1). I have found that stock cultures of *Ophiobolus graminis* are particularly liable to loss of pathogenicity in this way; no doubt many other mycologists have similarly lost both time and temper with other fungi. This difficulty in the maintenance of pathogenic wild-types has been much reduced by the modern practice of maintaining stock cultures in a dormant condition; agar cultures once grown can be covered with mineral oil or freeze-dried, or cultures can be grown on autoclaved soil that is then allowed to become air-dry.

Competition on Staled Nutrient Agar, and Effects of Surface Sterilants

The conclusions concerning saprophytic colonization of virgin nutrient agar do not apply to colonization of the same agar when it has become staled by fungal growth, as we have demonstrated in a recent study at Cambridge (Dwivedi & Garrett, 1968). We found that the species-spectrum of fungi colonizing staled agar changed progressively with the time and degree of staling due to fungal growth; the most highly staled agar was colonized only by species of *Penicillium* and *Trichoderma*, which showed a high degree of tolerance to fungistatic growth-products of other fungi.

When plating out pieces of plant tissue for assay of their fungal colonizers, the type of surface-sterilizing agent employed and the duration of the treatment exert a selective effect on the species-spectrum of fungi forming colonies on the isolation medium. This was shown by Walker's (1941) study of fungi colonizing wheat straw baits buried in various soils collected from fields at the Rothamsted Experimental Station. *Fusarium roseum* f. *cerealis* was the dominant fungus on the isolation plates after treatment of the straws with calcium hypochlorite, a mild sterilant, or after prolonged washing with sterile water. But after treatment of the straws with mercuric chloride or silver nitrate, the place of *F. roseum* was taken by species of *Penicillium* and *Trichoderma* which, after the longer periods of surface sterilization, were often the only fungi developing on the isolation plates.

This effect has been put to good use in the search for surface sterilants that are selective for particular fungal pathogens. In an early application of this principle, Davies (1935) found that a higher percentage isolation of *Ophiobolus graminis* from infected wheat tissue could be obtained by employing silver nitrate instead of the more usual mercuric chloride, and I have found that his method gives better results than any other I have tried. Since then, this principle has been widely employed in the search for selective surface sterilants, and for selective fungistatic agents to be incorporated in the isolation media designed for picking up particular pathogens.

COLONIZATION BY *RHIZOCTONIA SOLANI*

I have already discussed some of the evidence on the saprophytic behaviour of *R. solani* in Ch. 2 (p. 45), but have deferred further discussion until the latter part of the present chapter because interpretation of the more recent findings is greatly assisted by the work of Wastie (1961) on agar plate colonization, just described. I must first mention, however, some most elegant work by Bateman (1964) on the decomposition of cellulose by *R. solani* in pure culture. He demonstrated that *R. solani* could decompose cotton fibres, which are generally agreed to be the form of native cellulose most resistant of any to decomposition, and so this fungus must be regarded as a highly efficient cellulose-decomposer. Bateman's paper is accompanied by some excellent photomicrographs showing enzymic erosion of cotton and filter-paper fibres taken from his pure cultures of *R. solani* on these two substrates. He identified areas of cell-wall hydrolysis by their loss of birefringence under polarized light, thus recalling the earlier studies by F. Baker (1939) on cell-wall degradation in comminuted fragments of plant tissue by the rumen microfloras of various herbivores.

Much attention has been devoted to the saprophytic life of *R. solani* by G. C. Papavizas & C. B. Davey of the U.S.D.A. at Beltsville, Maryland. In their first papers on this subject (1959, 1961), they reported obtaining high percentages of *R. solani* from mature, dry stem segments of buckwheat (*Fagopyrum esculentum*) when buried in various soils for 4 days. Colonization of the baits by *R. solani* was determined by plating them out on water agar fortified with aureomycin, neomycin and streptomycin. If the baits were left in the soil for longer than 4 days, then percentage isolation of *R. solani* declined, presumably because of increasing competition for colonization of the agar from later fungal invaders of the bait tissues. Water agar is often employed instead of nutrient agar for attempted isolation of particular fungi from infected or colonized plant tissue, for the reason that both density of mycelial growth and production of fungistatic growth-products are restricted by the absence of nutrients, and particularly of carbon nutrients. But in view of Wastie's (1961) conclusion that success in fungal colonization of the virgin nutrient agar plate is determined chiefly by fungal growth rate, it is doubtful whether the precaution of using water agar in place of nutrient agar was of much help to Papavizas & Davey in isolating *R. solani* from their bait segments. For the particular case of *R. solani*, Wastie's data strongly suggest that its lack of success in colonization of the nutrient agar plate is due to the fact that it is not a fast grower, and even more to the fact that it is probably a late starter in the race. Thus out of eleven fungi for which full data were available, *R. solani* was ninth in percentage of colonies established on the plates, eighth in its growth rate on virgin agar, and first in its degree of tolerance to fungistatic microbial growth-products. I can further confirm this from my own early experience in attempting to isolate *R. solani* from roots of young cereal plants suffering from the 'no growth' disease in South Australia (Samuel & Garrett, 1932); it was rarely possible to isolate the pathogen from diseased roots unless they were in the early stage of infection.

Papavizas & Davey obtained higher percentages of isolation of *R. solani* when their baits were buried at soil moisture contents below rather than above 50% saturation. Rather surprisingly, the dead buckwheat stem segments gave higher percentages of *R. solani* on plating out than did infected hypocotyls of snap bean (*Phaseolus vulgaris*). In a subsequent paper, Papavizas & Davey (1962b) described some modifications in their plating out procedure for isolation of *R. solani* from mature stem segments (5–8 mm long) of twelve plant species tested as baits; the best results were given by buckwheat, cotton and lima bean and the poorest by maize, oats and soybean. Twenty clones of *R. solani* obtained from ten different soils were tested for pathogenicity against seedlings of snap bean, radish, soybean, sugar beet and wheat, and these saprophytically isolated clones were found to range from non-pathogenic to highly pathogenic; three clones were highly pathogenic to all five seedling species. Percentage isolation of *R. solani* from these baits ranged from 1% from a soil under grass for 3 years up to 97% from a soil cropped intensively with susceptible species for at least 6 years. Following upon a demonstration by Papavizas & Davey (1962a) that percentage saprophytic colonization of buckwheat baits by *R. solani* was progressively inhibited in four different soils by concentrations of carbon dioxide ranging from 10 through 20 to 30%, Papavizas (1964) studied carbon dioxide tolerance of twenty isolates each of *R. solani* and *Rhizoctonia praticola* Kotila, respectively. With a few exceptions, a high degree of competitive saprophytic ability amongst these forty isolates was found to be correlated with good tolerance of the higher concentrations of carbon dioxide. The results of Papavizas & Davey with stem segments as baits for *R. solani* have been confirmed by Zarka (1963), employing stem segments of *Corchorus olitorius* as substrate units in conjunction with the full Cambridge method; in his tests, the saprophytic performance of *R. solani* was as good as that we obtained with *Fusarium roseum* and *Curvularia ramosa* in colonizing wheat straw at Cambridge.

Nevertheless, it is unsafe to generalize about the relative competitive saprophytic abilities of different fungal species from a comparison between single isolates of each; this has been demonstrated by the variation in saprophytic performance amongst the forty isolates of *R. solani* and *R. praticola* in the experiments by Papavizas (1964) just quoted, and has been further emphasized for *R. solani* by Baker, Flentje, Olsen & Stretton (1967).

The findings of Papavizas & Davey in the U.S.A. have been very thoroughly confirmed by Sneh, Katan, Henis & Wahl (1966) in Israel, in a paper that provides some comparisons of particular relevance to this discussion. Sneh *et al.* employed the method of Papavizas & Davey, and their isolation medium for plating the baits was water agar + 250 ppm chloramphenicol. As baits they employed stem segments taken from plants of bean (*Phaseolus vulgaris*), cotton, buckwheat, oats, wheat and barley; the bean and cotton plants were grown to an age of 6 weeks, whereas stems of the other four species of crop plant were described as 'mature'. No one acquainted with these six plant species would doubt that the greatest contrast in condition of the plant tissues would have been provided by the living green tissues of the 6-weeks-old bean and cotton plants, on the one hand, and the mature, dry and nearly dead tissues of wheat and barley on the other. We should therefore expect that the green tissues of bean and cotton would possess at least some residual resistance to invasion, thus favouring parasitic fungi like *R. solani* over obligate saprophytes, whereas the nearly dead mature tissues of wheat and barley would have lost any significant resistance to invasion by obligate saprophytes. Sneh *et al.* have obligingly furnished us with a means of testing this assumption, by killing half their bait segments in the autoclave before burial in the soil. A relevant selection from their data is given in Table 8.

We can see in Table 8 that percentage isolation of *Rhizoctonia* species (*sic*) from the bait segments has reached a higher maximum, and has continued at a higher level, in the untreated *living* stem segments of bean and cotton than in those *killed* by autoclaving before burial in the soil. This is exactly the result we should have expected; the living segments retain enough tissue-resistance to

Table 8. *Percentages of untreated and autoclaved stem segments giving colonies of* Rhizoctonia *species on water-chloramphenicol agar after burial in a naturally infested soil for the periods shown*

(From data of Sneh *et al.* 1966)

		No. days burial in soil					
		2	3	4	7	14	21
Bean,	untreated	27	57	75	42	45	36
	autoclaved	46	57	56	27	18	25
Cotton,	untreated	24	64	70	66	45	45
	autoclaved	70	60	61	41	36	29
Barley,	untreated	51	47	38	7	0	0
	autoclaved	51	51	37	2	0	0
Wheat,	untreated	41	49	30	7	2	0
	autoclaved	50	51	30	6	0	0

delay colonization by competing obligate saprophytes but not sufficient to delay infection by the parasitic *Rhizoctonia* species. Killing the stem segments by autoclaving has destroyed the advantage enjoyed by *Rhizoctonia* species of being able to invade the plant tissues in advance of obligate saprophytes. When we look at the corresponding figures for stem segments of wheat and barley, we see that killing the tissues by autoclaving has made no significant difference to percentage recovery of *Rhizoctonia* species on the isolation plates; it is clear that these mature tissues must have been virtually dead before autoclaving, so far as any resistance to colonization by obligate saprophytes was concerned.

This experiment by Sneh *et al.* has been most admirably designed for a demonstration of this important distinction between infection of still living tissues by parasites and saprophytic colonization of dead tissues by obligate saprophytes. The results of this experiment, more convincingly than any others that I have seen, underline the importance of the practical recommendation that I made in Ch. 2 (p. 48); in any situation of disease risk, it is safer to employ dry, dead crop residues as an organic manure rather than the still living green tissues of a cover crop.

COLONIZATION OF WOODY SUBSTRATES

The problems I have just been discussing have not been easy ones to investigate; nevertheless, if we now pass on to consider competitive saprophytic colonization of woody tissues, as provided by the old roots, branches and trunks of bushes and trees, we encounter situations of still greater difficulty. We have just been considering the importance of distinguishing between infection of still living tissues by parasites, and colonization of dead tissues by obligate saprophytes; this can be a difficult distinction to make in practice for the tissues of herbaceous plants, but with woody perennials the difficulty is formidable indeed. Yet some way of making this distinction has to be found, because reliable root-disease control in horticultural and sylvicultural practice can be securely based only on a thorough understanding of the fundamental issues involved.

In Ch. 1 (p. 8), I referred briefly to the ring-barking control method introduced by R. Leach (1937, 1939) in East Africa against *Armillaria mellea*. By this method, old trees that have to be removed in preparation for establishment of a young plantation crop on the site are ring-barked a year or more before felling; the relatively early death of the root system that follows ring-barking permits earlier invasion of the tissues by saprophytic wood-decomposing fungi, and their occupancy brings to a halt further invasion by *A. mellea*. Prior saprophytic colonization thus provides a very useful control measure against the further spread of infection by specialized parasites like *A. mellea*, and so we need more information about this subject.

In an attempt to provide some information on this general question, I performed some laboratory experiments with *A. mellea* (Garrett, 1960*b*) of a similar type to that by Sneh *et al.* (1966) just described. As a substrate for *A. mellea*, I employed segments of willow (*Salix alba* var. *coerulea*) shoots (approx. 1·75 cm diam.), which were buried in two types of soil in laboratory containers. Before burial, half of the living shoot segments were killed by autoclaving. Segments were buried in soil for various periods up to an effective maximum of 49 days before inoculation with *A. mellea* rhizomorphs growing from woody inocula. The relative volume of each substrate segment colonized by the fungus after an incubation period of 4 weeks was then assessed by measuring weekly growth increments of rhizomorphs put out from the segment in soil growth tubes over an assessment period of 5 weeks, employing a technique devised in an earlier study (Garrett, 1956*b*). The substrate value of the living segments for *A. mellea* was maintained for a period of 7 weeks previous burial in soil, which was the maximum tested. But the substrate value of dead segments declined with increasing length of burial in soil from 3 weeks onwards; this I attributed to partial occupation of the segments by saprophytic fungi before they were inoculated with *A. mellea*. My results thus agreed with prediction from Leach's original hypothesis, though I had reduced the time-scale as well as the space dimensions of the experiment from those of the plantation crop to those of the laboratory.

The results of this experiment thus suggest that *A. mellea* can colonize dead woody tissues in the soil, or above it, if favourably placed to do so. Findlay (1951) has reported extensive damage by *A. mellea* to pit props in very damp coal mines. Other fungi infecting tree roots may similarly have opportunities for purely saprophytic invasion of woody tissues. Wallis & Reynolds (1965) have demonstrated that this is at least a possibility for *Poria weirii* Murr., causing a root disease of Douglas fir. Stakes of Douglas fir heartwood, air-dried for 6 months and thus 'presumed dead', were buried beneath a 50-year-old stand of Douglas fir in British Columbia for periods of 0, 3, 6 and 12 months, respectively. Ten replicate stakes at each sampling period were removed from the soil, inoculated with wood-block inocula of *P. weirii* and then reburied. Saprophytic colonization of these stakes by *P. weirii* from the inocula did not decline with increasing periods of burial in the soil up to 6 months; after 12 months, however, there was a significant reduction in the extent of saprophytic colonization by *P. weirii*.

6   Saprophytic survival of root-infecting fungi in infected or colonized host tissues

Saprophytic survival is defined as mycelial survival of a root-infecting fungus in infected host tissues invaded during its parasitic phase. We can also include in this term the mycelial survival of a parasite in plant tissues of any kind that it has been able to colonize competitively in its saprophytic phase, if it has one. In respect of saprophytic survival, no valid distinction can be made between the mechanism of mycelial survival in infected and saprophytically colonized tissues, respectively; in fact, most of the investigations into saprophytic survival in infected tissues have actually been carried out with colonized tissues. For such experiments, large numbers of infected or colonized tissue-units are required; infected material of uniform quality is difficult to collect from diseased crops in the field in sufficient quantity for a large experiment, and laborious and time-consuming to produce from inoculated plants in the glasshouse or in small field plots. But by inoculating prepared tissue-units after autoclaving in culture flasks, large quantities of material uniformly colonized by the desired pathogen can be prepared with the minimum of labour. Confirmatory tests have shown that such colonized tissue-units satisfactorily reproduce the behaviour of naturally infected tissue-units so far as saprophytic survival of the pathogen is concerned.

During infection of living tissues or saprophytic colonization of dead tissues, the fungus establishes a mycelial network throughout the tissues. Thereafter, whether the dead tissues remain in soil or are cultivated into it, the mycelial network of the pathogen survives saprophytically by a slow but continuous decomposition and exploitation of the tissues; sugars and the more easily assimilated compounds are quickly exhausted but the cellular framework of hemicellulose, cellulose and lignin is more resistant to decomposition, and provides for a considerable longevity of the pathogen;

this longevity extends up to a year and sometimes more for pathogens of annual crops and for periods of years for pathogens of woody perennials. As I have remarked above, survival of a pathogen in inoculated and colonized tissue-units seems to reproduce satisfactorily, for experimental purposes, its survival in similar units of naturally infected tissue. In either case, the pathogen establishes its mycelial network in the plant tissue well in advance of other invaders; during infection, the pathogen invades living plant tissue, and host resistance keeps out secondary fungi, at least until the mycelium of the pathogen has completely occupied most, if not all, of the host tissue. Inoculation of sterilized tissue-units with the pathogen reproduces, reasonably well according to all available evidence, infection of living tissues for the production of survival-units. But when an unspecialized parasite with sufficient competitive saprophytic ability colonizes competitively dead crop residues that have been cultivated into the soil, it may not achieve so complete an occupation of the plant tissues as if it had infected them during its parasite phase; such incomplete occupation of the plant tissues is likely to be reflected in a decreased longevity of saprophytic survival.

Saprophytic survival of the mycelial network in plant tissues needs to be sharply distinguished from *dormant survival* of a pathogen, within the tissues or set free in the soil, in the form of resting spores or sclerotia (Ch. 7). Such a distinction is easy to make in theory, but special precautions may have to be devised in experimental practice to separate these two types of survival, when they occur concurrently during the survival of a single fungus in plant tissues.

For detailed discussion of the experimental evidence on saprophytic survival, I shall follow my precedent of the preceding chapter and separate host plants into herbaceous annuals and woody perennials.

SURVIVAL IN TISSUES OF HERBACEOUS HOSTS

Most of the investigations so far conducted into saprophytic survival have been made with the five cereal foot-rot fungi discussed in Ch. 5, viz. *Ophiobolus graminis, Fusarium roseum* f. *cerealis* (syn. *F. culmorum*), *Cochliobolus sativus* (stat. conid. = *Helminthosporium sativum*), *Curvularia ramosa,* and *Cercosporella herpotrichoides.* The specialized ectotrophic parasite, *O. graminis,* infects first the root system, and thence the stem (tiller) bases of its cereal hosts; the root-rot thus develops into a foot-rot. There is less certainty about the usual sequence of infection in the foot-rots caused by *F. roseum, C. sativus* and *C. ramosa* but it seems probable that stem-base tissues are often invaded by these less specialized pathogens direct from the surrounding soil, and not necessarily via infection of the roots. The specialized pathogen *C. herpotrichoides,* causing the eyespot lodging disease of cereals, is not a root-infecting fungus at all. Primary infection of autumn-sown cereal seedlings occurs during the winter and early spring from rain-splashed spores; these spores are abundantly produced, when night temperatures are around 0° C, on infected stubble from a previous diseased cereal crop that has been incompletely buried during ploughing (Glynne, 1965). Although *C. herpotrichoides* does not infect the roots of its cereal hosts, the fungus is a soil-borne pathogen inasmuch as it survives saprophytically in infected cereal stubble and so its survival has been studied by the same techniques as have been employed for the other four fungi listed above. To these five fungi has recently been added a sixth; this is *Cephalosporium gramineum* Nisikado & Ikata (perfect state = *Hymenula cerealis* Ellis & Everh.), which causes a vascular wilt, known as stripe disease, of winter wheat in Washington State, U.S.A. (Bruehl, Lai & Huisman, 1964).

In my own early studies of the saprophytic survival of *Ophiobolus graminis,* I employed a standardized survival-unit of wheat straw tissue colonized in pure culture by the fungus, though I took the precaution of confirming the more important conclusions resulting from such experiments by comparable tests with pieces of naturally infected wheat straw collected from diseased crops around harvest time (Garrett, 1938, 1940, 1944). These survival-units were cut from lengths of mature wheat straw to give pieces about 4 cm long, each with a node at the lower end, so that the internodal part above it consisted of the culm surrounded by the subtending leaf sheath; inclusion of a node with each straw piece was essential to secure coherence during decomposition in the soil for prolonged periods, and during subsequent manipulation in the test for survival of viable *O. graminis*. After the straws had been colonized in pure culture by *O. graminis*, during incubation for about one month at 25° C, they were buried in lots of fifty in approx. 200 ml soil in glass tumblers or jars. At intervals, a sample of not less than 100 straws in two soil containers was taken for the survival test; a wheat seed was planted, germ end downwards, within the lumen of each straw, and fifty straws thus inseminated were planted in a seedling box filled with sand, which is a rooting medium optimum for infection and spread by *O. graminis* (Garrett, 1936). This wheat seedling test is the only reliable one for *O. graminis*, which is a poor competitor on the nutrient agar plate (Ch. 5); the microconidia of this fungus are unreliable for diagnosis, and production of perithecia in a moist chamber cannot be induced either quickly or with the complete assurance required for a survival test. This difficulty does not arise with *Cochliobolus sativus* or with *Curvularia ramosa*, which produce their large and distinctive, septate, dark-coloured conidia as soon as straws are incubated in a moist chamber of the sand-plate type devised by Butler (1953 *a*) and employed by him in subsequent studies of saprophytic survival (Butler, 1953 *c*, 1959). When straws have been incubated on a sand plate for about 1 week at 25° C, sporulating conidiophores on the straw surfaces can be recorded during rapid scanning under the stereoscopic microscope; if a detailed record is required, the complete length of the straw can be divided up into successive fields of view, each of 1 mm diam., by employing a mechanical stage. Survival of *Fusarium roseum* can be reliably assessed by plating out surface-sterilized straws on nutrient agar, because *F. roseum* has a high competitive saprophytic ability for this substrate (Butler, 1953 *a*; Wastie, 1961). As for *Ophiobolus graminis*, survival of *F. roseum*, *C. sativus* and *C. ramosa* can be assessed by my wheat seedling test; the test is sufficiently reliable for experiments in which straws have been completely colonized by the original inoculant fungus, and there is no substantial probability that the test wheat seedlings will be infected by other soil-borne fungal pathogens. This proviso has to be made, because roots infected by all three pathogens show a generalized discoloration due to cortical necrosis, and symptoms of root infection by one pathogen cannot be distinguished from those produced by the other two. This difficulty of recognition does not apply to roots infected by *Ophiobolus graminis*, which show two distinctive features: (1) an intense necrosis and discoloration of the vascular cylinder; (2) external growth of the characteristic, dark-coloured runner hyphae, which are easily visible under the stereoscopic microscope. The wheat seedling test for survival of *F. roseum* can be replaced either by plating out surface-sterilized straws on nutrient agar, as noted above, or by a modification of Butler's sand-plate test introduced by Macer (1961 a). The hyaline macroconidia of *F. roseum* are neither large nor distinctive enough to be identified with certainty under the stereoscopic microscope but they can be scraped off the straw surface into a drop of water, which is then searched under a sufficiently high power of the compound microscope. Macer employed this modified sand-plate method also for assessing survival in straw of *Cercosporella herpotrichoides*; to induce sporulation by this fungus, sand plates have to be subjected to a daily alternation of temperature with night temperature going down to near o° C, as was first demonstrated by Glynne (1953). For *C. herpotrichoides*, Macer also successfully employed a wheat seedling test; nodes were cut off from straws after recovery from the soil, and the internodal segments were slipped over the coleoptiles of just emerged wheat seedlings as a collar. The inoculum segments were then covered with a layer of moist sand to ensure infection, the occurrence of which was demonstrated after some 8 weeks by appearance of the characteristic eyespot lesions.

During the first part of my investigation into the saprophytic survival of *O. graminis*, it became possible to correlate the rapidity of disappearance of the fungus from the straws in a general way with the maintenance of soil conditions optimum for microbial activity; thus loss of viability was most rapid under conditions of medium to high temperature, suitable moisture content, and good soil aeration. Viability of *O. graminis* was preserved, on the other hand, at a low temperature (2–3° C) and in air-dry soil, and was actually maintained for longer in a water-logged soil than in one held at a moisture content of 50% saturation. The general interpretation placed upon these results is that under conditions conducive to maximum microbiological activity, the available food reserves of the straw substrate are most quickly exhausted by *O. graminis* itself and by associated micro-organisms that have followed it into the tissue. Conditions restricting microbiological activity, such as low temperature, dryness and poor aeration, are thus likely to promote maximum longevity of the fungus in infected host tissues, providing that the effect of such conditions is not so extreme as to be actually lethal. Thus *O. graminis* is likely to suffer less loss of viability during the low temperatures of winter, or during the extreme drought of such a summer as occurs in South Australia, than during periods of the year in which soil temperature and moisture content are more favourable for microbiological activity.

This conclusion has been fully confirmed by the work of Fellows (1941), using a technique approximating more closely to actual survival of *O. graminis* in the field. Fellows studied the rate of disappearance of *O. graminis* in naturally infested soil taken straight from the field and incubated under the eight possible combinations of high and low temperature, loose and firm packing, and moist and air-dry conditions. The low-temperature samples were stored in a deep cave and the high-temperature ones were kept in the glasshouse. Looseness was maintained by occasional stirring, and a firm packing achieved by thorough tamping of the soil whilst moist. The moist soils were maintained at approximately 60% of saturation, whereas the others were allowed to air-dry. Four experiments were made, and periods of incubation varied from 230 to 777 days; at the end of the incubation period, all soil samples were brought to favourable tilth and moisture content, and a test crop of wheat grown in them to maturity, in the glasshouse. The degree of infection on the plants at harvest was expressed by means of a 'severity rating'. In almost every comparison provided by this series of extensive factorial experiments, the fungus disappeared more quickly in a warm soil than in a cold one, and more quickly in a loose soil than in a compact soil. The influence of moisture was less straightforward; under cool conditions survival appeared to be rather better in a moist than in a dry soil, whereas under warm conditions the reverse was the case. In all four series of experiments, without exception, disappearance of the fungus was most rapid in the warm, moist, loose soil, i.e. under just those conditions of high temperature, adequate moisture content and good aeration postulated above as most conducive to rapid loss of viability.

Effect of nitrogen on saprophytic survival

The C/N ratio of mature wheat straw varies around a value of 80; mature stem tissues of other annual crop plants similarly have a fairly high C/N ratio. It was first realized many years ago, as recounted by Waksman (1932), that such mature plant tissues when composted were decomposed much more quickly if nitrogen, in some readily available form, was added to the compost; similarly, tissues of this type disappeared more rapidly in nitrogen-rich than in nitrogen-poor soils. The mechanism of this effect depends upon the fact that microbial degradation of pectins, hemicelluloses, celluloses and lignins of the plant cell wall is brought about by the action of enzymes. These enzymes are both bound to the microbial cell wall and set free in soil moisture films; the action of bound enzymes is necessarily very local and that of the free enzymes is limited by the number of enzyme molecules produced and by the effective radius of their diffusion. The largest constituent of mature plant tissue is cellulose and it is this, together with the superimposed lignin, that survives for longest in the decomposing tissue. From an active length of fungal hypha lying alongside a plane of plant cell wall, cellulase enzymes diffuse outwards and the products of cellulose hydrolysis, i.e. cellobiose and eventually glucose, diffuse inwards so that a proportion is absorbed by the fungal hypha. So the radius of cellulase action around the enzyme-producing fungal hypha must be limited both by the intensity and duration of its production, and also by any factors limiting its outwards diffusion. For these various reasons, no individual fungal hypha can produce more than a limited zone of enzymic erosion around itself; when this zone is exhausted, then this particular fungal hypha will die from carbon starvation.

Zones of enzymic erosion around fungal hyphae and bacterial cells have been demonstrated under the microscope by the use of differential staining methods and also by loss of birefringence of degraded cellulose when viewed by plane-polarized light. Both methods were used by F. Baker (1939), who has figured very clearly these zones of erosion around bacterial cells lying alongside cell walls in fragments of plant tissue taken from the alimentary canal of various herbivorous animals. Bateman (1964) has similarly figured them on cotton and filter-paper fibres taken from his pure cultures of *Rhizoctonia solani* on these two types of native cellulose. Enzymic erosion is also demonstrated particularly well by the so-called 'boreholes' around hyphae of wood-destroying fungi; Bailey & Vestal (1937) have used the geometrical configuration of these boreholes as an aid to interpreting the fine structure of cellulose fibrils. Proctor (1941) has provided microscopical evidence to show that penetration of the thickened secondary walls of plant xylem elements by fungal hyphae is through enzymic action and that mechanical pressure plays only a subsidiary role at most; he has suggested that the term 'borehole' is thus a misnomer. The observed fact that these boreholes do not usually attain a diameter exceeding three times that of the fungal hypha that has made them (Cowling, 1963) suggests that the functional life of any particular length of fungal hypha in cellulolysis is of limited duration. In decomposing cellulosic tissues of herbaceous plants, which are of a construction much looser than that of secondary wood in woody perennials, there must be a less effective barrier to outwards diffusion of both cellulase and the sugars derived from its action. The sugars will be quickly absorbed by associated micro-organisms living as commensals of the cellulose-decomposer in the plant tissue (Ch. 5, p. 116). We can therefore conclude that production of cellulase by a fungal hypha is likely to have become limited by the operation of natural selection to an optimum or economic quantity; we can certainly visualize the possibility that production of cellulase in excess of this optimum

quantity would be uneconomic, because further increases in production would yield a progressively diminishing return of sugar
by inwards diffusion towards the cellulase-producing hypha in the
presence of sugar-absorbing commensal organisms. A way in
which this limitation of cellulase-production may be brought about
is suggested by the fact that the cellulases are *adaptive* (as distinct
from *constitutive*) enzymes and so contact with cellulose is required
for their induction; hydrolysis of the immediately surrounding
cellulose will remove a fungal hypha from contact with its cellulose
substrate, and so induction of cellulase will cease.

Nevertheless the mycelium of which our particular length of
hypha under consideration forms a part can survive granted one
essential condition; this is that a continued slow growth and
extension of the mycelium within the substrate tissue is not
limited by lack of nitrogen or of any other mineral nutrient. So
we can now understand that, because both the working life and the
survival of individual hyphae are limited by the radius of enzyme
action, it must follow that a continued extension of the mycelium
by branching and growth within the substrate tissue is an essential
condition for continued cellulolysis, and also for survival of the
mycelium effecting that cellulolysis. Rate of decomposition of
plant tissue is thus limited by the supply of mineral nutrients
essential for the synthesis of fungal and other microbial protoplasm.
Nitrogen is the mineral element required in largest amount,
followed by phosphate and potash. Decomposition-rate of mature
plant tissue in compost or in nitrogen-poor soil is thus limited
by the supply of available nitrogen; phosphate may sometimes
limit decomposition-rate in compost, but much more rarely in the
soil.

An understanding of the mechanism of microbial decomposition
of plant tissue, as outlined above, is an absolute prerequisite for
comprehension of the factors determining longevity of saprophytic
survival by root-infecting fungi. At the outset of my work on
saprophytic survival of *Ophiobolus graminis*, I had anticipated that
rate of its disappearance from the wheat straws would be correlated with the rate of overall microbial activity in the soil. In
general terms this turned out to be true, as I have outlined above,

but with respect to the effect of nitrogen my original expectation was exactly reversed. *O. graminis* was found to die out early from the rigid and little-decomposed straws buried in nitrogen-poor soils, whereas it was found to survive in almost 100% of straws in nitrogen-rich soils that had become so soft, attenuated and disintegrated by decomposition that they were difficult to handle for the wheat-seedling test. At first I adopted a wrong hypothesis to explain this quite unexpected finding (Garrett, 1938), but further thought upon the problem followed by microscopical examination of the straws after burial in the soil showed me that the mycelium of *O. graminis* continued to develop in the straw tissue lying within the soil, and that this development was greater, and continued for longer, in nitrogen-rich soil (Garrett, 1940). My observations have recently been confirmed by Chambers & Flentje (1969). We can therefore conclude that *O. graminis* dies out prematurely from carbon starvation in wheat straw buried in nitrogen-poor soil, because it cannot obtain the nitrogen needed for continued mycelial growth and hydrolysis of fresh areas of cellulose within the straw tissue. The fungus is thus starving in the middle of a well-stocked larder, because it can no longer get at its food.

After this work on *O. graminis*, I had fully expected that we should find a similar effect of nitrogen in prolonging saprophytic survival of other cereal foot-rot fungi, and perhaps also of other root-infecting fungi similarly surviving in other mature plant tissues with a high C/N ratio. We were therefore surprised when Butler (1953c) discovered that excess of soil nitrogen actually reduced longevity of *Cochliobolus sativus* in wheat straw. Subsequently Butler (1959), continuing this work in New South Wales, confirmed this result by comparing longevity of saprophytic survival of *C. sativus* and three other cereal foot-rot fungi in a ley soil of high fertility (total N = 0·13%) and in an arable soil of low fertility (total N = 0·07%). Survival of both *O. graminis* and *Fusarium roseum* f. *cerealis* was longer in the high-nitrogen soil, whereas that of *C. sativus* was longer in the low-nitrogen soil; survival of *Curvularia ramosa* appeared to be indifferent to the level of soil fertility and nitrogen content (Fig. 19).

In discussing these results, Butler suggested that this effect of

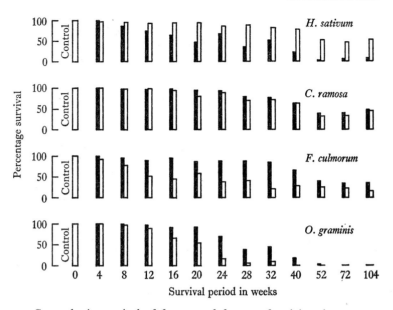

19. Saprophytic survival of four cereal foot-rot fungi in wheat straw. The left-hand column in each pair represents percentage survival in a soil of high fertility, and the right-hand one represents survival in a soil of low fertility. (After F. C. Butler, 1959.)

nitrogen in actually reducing longevity of *Cochliobolus sativus* might be connected with its low tolerance of fungistatic growth-products excreted by competing soil micro-organisms (Butler, 1953*b*; Wastie, 1961). He supposed that this might limit the slow but continued mycelial development such as I had observed for *O. graminis* in the wheat straw tissue; if further mycelial develop-ment of *C. sativus* were thus limited, then diffusion of soluble nitrogen into the straw tissue from the surrounding soil would not benefit *C. sativus*, but might well help other soil micro-organisms, and especially soil fungi, to invade the substrate, and thus to increase the population of competing micro-organisms within the straw tissue. At this stage of the discussion, it will be helpful to summarize evidence on effects of nitrogen on longevity of sapro-phytic survival by five cereal foot-rot fungi; in Table 9, a 'positive effect' of nitrogen indicates that this nutrient prolongs survival, whereas a 'negative effect' denotes the reverse.

Table 9. *Effects of nitrogen on longevity of saprophytic survival by five cereal foot-rot fungi*

Pathogen	Disease	Effect of nitrogen	Authors
Ophiobolus graminis	Take-all	Strongly positive	Garret (1938, 1940, 1944) Butler (1953*c*, 1959) Macer (1961*b*) Chambers & Flentje (1969)
Fusarium roseum f. *cerealis*	Foot-rot	Strongly positive	Butler (1959)
Cercosporella herpotrichoides	Eyespot	Weakly positive	Macer (1961*b*)
Curvularia ramosa	Foot-rot	Weakly positive Indifferent	Butler (1953*c*) Butler (1959)
Cochliobolus sativus	Foot-rot	Strongly negative	Butler (1953*c*, 1959) Garrett (1966*b*)

We can see in Table 9 that a strong positive effect of nitrogen on longevity of *O. graminis* has been repeatedly confirmed and so has a strong negative effect on longevity of *C. sativus*. The strong positive effect on *F. roseum* rests on the results of a single experiment by Butler (1959). Responses to nitrogen by *C. ramosa* and *C. herpotrichoides*, respectively, have been either weak or indifferent.

I suspected that these differing responses to nitrogen in longevity of saprophytic survival by these five fungi might be connected in some way with their respective cellulolytic abilities. The cellulose content of mature wheat straw comprises about 40% of total dry matter, yet no information on cellulolytic ability of these five species appeared to be available. To procure this, I measured loss in dry weight by filter-paper cultures of the five fungi over a standard incubation period of 7 weeks at 22·5° C. *C. sativus* was found to have the highest cellulolysis rate and it seemed possible that this was somehow connected with its strong negative response to nitrogen in longevity of survival. Nevertheless, cellulolysis rate by itself could not be correlated at all clearly with the nitrogen effects; I then came to realize that the adequacy of the cellulolysis rate for continued saprophytic survival of a fungus in wheat straw

Table 10. *Cellulolysis adequacy indices for five cereal foot-rot fungi*

	No. isolates tested	Percent. loss in filter-paper dry wt. after 7 wks at 22·5° C		Linear growth rate of fungus (mm/24h) at 22·5° C	Mean value for cellulolysis adequacy index
		Range	Mean		
Ophiobolus graminis	5	2·1–3·6	2·8	7·3	0·38
Fusarium roseum f. *cerealis*	4	6·5–8·5	7·5	11·0	0·68
Cercosporella herpotrichoides	4	0·8–1·1	0·9	1·3	0·69
Curvularia ramosa	1	—	6·6	5·3	1·25
Cochliobolus sativus	6	5·4–8·4	7·5	3·6	2·10

tissue is not determined by this rate alone, but by the relation between this rate and the general metabolic rate of the fungus. The general metabolic rate determines the rate at which carbohydrates are consumed by the processes of respiration and growth; the cellulolysis rate determines the rate at which soluble carbohydrates are made available for the metabolism of the fungus. Only if the cellulolysis rate is *adequate* to maintain the general metabolic rate will a fungus be able to continue surviving in the straw tissue. Thus a fast-growing fungus with a high metabolic rate will require a higher rate of cellulolysis to supply its needs than will a slow-growing fungus with a low metabolic rate. To express these relationships, I proposed the *Cellulolysis Adequacy Index*, which is obtained by dividing cellulolysis rate by linear growth rate (as a parameter of general metabolic rate), each expressed in suitable units. If cellulolysis rate is expressed by percentage loss in dry weight of filter-paper cultures over the standard incubation period of 7 weeks at 22·5° C and linear growth rate as mm/24 h at 22·5° C, then the cellulolysis adequacy index is found to vary around unity for the five cereal foot-rot fungi under investigation. In Table 10, mean values of the index for *F. roseum, C. herpotrichoides* and *C. ramosa* have been taken from Garrett (1966b) and the mean value for *O. graminis* from Garrett (1967); that for *C. sativus* is taken from my hitherto unpublished data (Table 10).

A comparison of the data given in Table 10 with those in

Table 9 suggests that a low value of the cellulolysis adequacy index (hereafter referred to as the CAI) is associated with a strongly positive response to nitrogen in longevity of survival and a high value of the index with a strongly negative response. The concept of the CAI can thus explain why a fungus such as *Cochliobolus sativus* does not need extra nitrogen in order to survive in wheat straw tissue; a further extension of this hypothesis can also explain why extra nitrogen actually *shortens* its period of survival. This is because a supply of nitrogen much in excess of the level needed to keep cellulolysis going at a rate just adequate for survival is going to promote an *uneconomic* rate of cellulose consumption; this will result in a complete exhaustion of cellulose in the substrate in a shorter time than if nitrogen had been supplied at a rate just sufficient to keep the fungal colony alive and metabolizing the substrate.

This hypothesis thus postulates that the primary factor determining the response of a fungus to nitrogen is its own behaviour in metabolizing the cellulose of the substrate, as expressed by its CAI value, and that other fungi and other soil micro-organisms play a subsidiary role in decomposition and eventual exhaustion of the substrate. The *direct effect* of nitrogen upon the cellulolytic activity of the fungus itself thus differs from the *indirect effect* proposed by Butler (1953b), who thought that nitrogen acted by assisting other micro-organisms to invade and colonize the substrate alongside its primary colonizer. These two effects are not mutually exclusive but I designed an experiment with *Cochliobolus sativus* to determine which effect was the more important (Garrett, 1966b). A total amount of nitrate nitrogen sufficient to curtail saprophytic survival of *C. sativus* was provided in two ways: (1) to the straws in the pure culture flasks before inoculation; (2) to the soil at time of burying the colonized straws. I argued that if the effect of nitrogen was mainly on the cellulolytic activity of *C. sativus* itself (the *direct effect* hypothesis), then nitrogen would reduce longevity more if added to the pure culture flasks, because it would have a month longer in which to act on the fungus, and because most of it would be absorbed and retained by the mycelium of *C. sativus*. But if the main effect of nitrogen was to assist other fungi and micro-organisms to invade and colonize the substrate (the *indirect effect* or *replacement* hypothesis), then nitrogen would reduce longevity of *C. sativus* more if it was added to the soil at time of straw burial. The longevity of *C. sativus* was estimated on a sample of fifty straws from one soil jar for each of the six experimental treatments (two fungal isolates times three nitrogen treatments) over the period 4–16 weeks. After recovery from the soil, the encasing leaf-sheath around each straw was removed and the surface of the culm segment was well rubbed under the running tap before sand-plating. The object of this was to remove any conidia, which act as survival spores in *C. sativus* (Ch. 7, p. 189); if this precaution had not been taken, dormant survival by conidia attached to the straws might have been wrongly recorded as saprophytic survival. After 9 days incubation on the sand plates at 22·5° C, each straw was thoroughly scanned under the stereoscopic microscope (50 × magnification); any straw bearing one or more erect, sporulating conidiophores was recorded as containing still viable *C. sativus* (Table 11).

Table 11. *Effect of nitrogen and its time of application on longevity of saprophytic survival by* Cochliobolus sativus *in wheat straws buried in soil*

(From Garrett, 1966b)

| | Percentage straws with *C. sativus* still viable | | | |
| | Period in soil (weeks) | | | |
	4	8	12	16
Isolate P1, no nitrogen	100	100	92	82
soil nitrogen	100	100	60	14
pure-culture nitrogen	100	72	48	4
Isolate BS790, no nitrogen	100	100	100	74
soil nitrogen	100	100	92	42
pure-culture nitrogen	100	100	78	2

The results of this experiment, displayed in Table 11, have returned a definite answer in favour of the *direct effect* as the more important of the two. After this, I realized that the most con-

Table 12. *Effect of nitrogen level on longevity of survival by* Ophiobolus graminis *in pure culture on filter-paper cellulose*

(From Garrett, 1967)

	Percentage roots infected by *O. graminis* Time of sampling (weeks)		
	10	15	23½
Nitrogen series S/4	80	80	16
S/2	99	97	55
S	98	99	92
2 S	100	85	82
4 S	100	73	73

Standard error of means = 3·79: least difference for significance at 5 % level = 11.

vincing evidence for the direct effect hypothesis would be to reproduce the whole spectrum of nitrogen responses in a fungus growing in pure culture upon a cellulose substrate. Although, over the range of nitrogen levels tested, a significantly negative response had been shown only by *C. sativus*, it seemed likely that the response of any of these five fungi would switch from positive to negative once the critical nitrogen level had been exceeded. The optimum nitrogen level for survival of *Ophiobolus graminis* in straw buried in soil is around 0·5 g N/100 g air-dry straw; results of an earlier experiment suggested that 0·9 g N was supra-optimal for maximum longevity of survival (Garrett, 1944, Table 1). *O. graminis* was selected as the subject for this experiment: (1) because its saprophytic survival is shorter than that of the other four fungi; (2) because symptoms of root infection in the wheat-seedling test are completely reliable for diagnosis. As most fungi survive in pure culture for a long time, I took all possible precautions to reduce the survival period in this experiment, and the substrate was reduced to a single circle of Whatman no. 3 filter paper lying on a reservoir of inorganic nutrient salts in pure quartz sand. At each sampling time, survival of *O. graminis* on the filter-paper circle was estimated by percentage infection of the roots of twelve wheat

158

seedlings planted thereupon; percentage figures given in Table 12 are based on examination of some 120 roots for each treatment-series at each sampling time (Table 12).

Table 12 shows that the optimum level of nitrate nitrogen for maximum longevity of *O. graminis* on filter-paper cellulose was that provided by series S, which was 0·235 g N/100 g air-dry filter paper. At the final sampling after 23½ weeks, the effect of the highest nitrogen treatment (0·94 g N/100 g air-dry filter paper in the 4S series) in reducing longevity of *O. graminis* was highly significant (at the 0·1 % level). The results of this experiment were thus in complete accord with prediction from the *direct effect* hypothesis concerning the various effects of nitrogen on longevity of saprophytic survival.

Other factors affecting survival

I have already mentioned the possibility that saprophytic survival of mycelium may be augmented by the production of dormant propagules in the host tissues. *Ophiobolus graminis* is not known to produce any type of dormant propagule; in the experiment with *Cochliobolus sativus* to which Table 11 refers, I had taken due precautions to separate saprophytic mycelial survival, which I was studying, from dormant survival by conidia. This precaution was necessary, because conidia of *C. sativus* function as dormant propagules in soil; their effectiveness as such is more readily understood since the recent announcement by Meronuck & Pepper (1968) that chlamydospores are produced within the cells of conidia overwintering in soil (Ch. 7, p. 190). Since this complication of saprophytic mycelial survival by dormant survival was eliminated from the comparison between *O. graminis* and *C. sativus* in the previous section, there seems no reason to doubt that the contrasting effects of nitrogen on the longevity of these two fungi in wheat straws were indeed produced by the influence of nitrogen on saprophytic mycelial survival. But this conclusion cannot be extended without qualification to the nitrogen response of *Fusarium roseum* f. *cerealis*, the only other one of these five cereal foot-rot fungi showing a strong response (positive) to nitrogen in its longevity of survival in wheat straws (Table 9). Cook (1968) has observed the regular production of mycelial chlamydospores by *F. roseum* in infected wheat tissues. We therefore have to ask ourselves whether nitrogen, which increases longevity of *F. roseum* in wheat straws, does so by promoting an increased production of chlamydospores in the infected or colonized tissues. To this question we can return a cautious negative; the available evidence suggests that production of chlamydospores by *Fusarium* species, as well as that of other types of dormant propagule, is favoured by a high rather than by a low C/N ratio of plant tissues (Ch. 7, p. 199 and elsewhere).

Although we can thus conclude that the effect of nitrogen on longevity of these cereal foot-rot fungi in wheat straw operates mainly through its influence on saprophytic mycelial behaviour, it seems most likely that overall longevity is determined by whether dormant propagules are produced and if so by their efficiency in promoting longevity. *Ophiobolus graminis* is the only one of these fungi that is not known to produce any type of dormant propagule, and it survives in wheat straw for not more than a year; Butler (1959) obtained survival of the other three fungi shown in Fig. 19 for a period of up to 2 years. The fifth fungus, *Cercosporella herpotrichoides*, was found by Macer (1961b) to survive under some soil conditions in approx. 75% of colonized straws buried outside in the field for 3 years. This difference in longevity between *O. graminis* and *C. herpotrichoides* agrees with the experience of English farming practice; the take-all disease can usually be controlled by a one-year break in the rotation between susceptible crops, whereas the eyespot lodging disease requires a break of several years. *C. herpotrichoides* was reported by Sprague (1931) and Dickens (1964), to produce micro-sclerotia on nutrient agar, and I also have observed these bodies, but their function and efficiency as dormant propagules in wheat straw buried in soil needs to be investigated.

In view of the intensity of microbial competition for substrates in the soil (Ch. 5), one of the most interesting facts brought to light by these and other studies on saprophytic survival is the remarkable longevity of pioneer fungal colonizers of plant tissue substrates lying in the soil. As we have just noted, *Ophiobolus graminis* survives under favourable soil conditions for up to a year in wheat straw, and its survival is not assisted by production of any dormant propagules; on the nutrient agar plate, at least, it is particularly sensitive to fungistatic growth products of other soil micro-organisms (Table 6, p. 132). How does this come about? Earlier in this chapter, I have emphasized the fact that when a primary parasite invades living plant tissues, host resistance to infection protects it, for a time at least, from following fungi, and so it establishes itself virtually in pure culture in the host tissues; in experiments with colonized plant tissues, this situation has been reproduced by establishing pure cultures of the inoculant fungi on sterilized host tissue. Establishment of the mycelium as a pure culture throughout the plant tissue must give the inoculant fungus a substantial initial advantage, but how is this advantage maintained when the tissue is buried in soil and open to invasion by all the surrounding soil micro-organisms? It has been suggested that production of fungistatic substances by the primary colonizer enables it to obstruct invasion of its substrate by following fungi, but this explanation could fit only a minority of the situations already investigated. Amongst the five cereal foot-rot fungi so far discussed, only *Cochliobolus sativus* is known to produce strongly fungistatic culture filtrates (Butler, 1953b), to which it is itself sensitive, as demonstrated by strong staling of cultures on most nutrient agars. Such a suggestion has been advanced by Tyner (1966), on the basis of his demonstration that *C. sativus* produces a fungistatic culture filtrate when grown on ground-up wheat straw. But demonstration of such an effect during early growth of the fungus on this substrate does not tell us whether the fungus is still producing a significant concentration of fungistatic substances in the later stages of its saprophytic survival, when it is subsisting on the relatively low concentrations of sugars released by cellulolysis, and when it is in greater danger than earlier of being replaced by other fungi from the surrounding soil. Antibiotics, as they are produced in cultures of micro-organisms both experimentally and industrially, are products of 'shunt metabolism', to employ an apt description coined by Foster (1949, pp. 164–9). Antibiotics are produced by fungi in pure culture on a sufficiently rich nutrient medium when mycelial growth has been halted by some limiting factor, such as a nutrient deficiency. In this situation, the mycelium is unable to stop absorbing sugar from the culture medium, and the excess sugar is excreted as a shunt metabolite, which may have antibiotic properties, as Woodruff (1966) has explained in an admirably lucid review of antibiotic production. These conditions for antibiotic production may sometimes be satisfied when a fungus first invades a substrate of fresh plant tissue, as I have already suggested in Ch. 5; for instance, Brian, Elson & Lowe (1956) have demonstrated production of patulin by *Penicillium expansum* Link em. Thom during its invasion of apple fruits, which have a fairly high sugar content. But it seems unlikely that the sugars released by cellulolysis of residual cellulose during the later stages of saprophytic mycelial survival in a mature plant tissue will reach a high enough concentration to permit a significant production of antibiotic. The rate of cellulolysis is self-limiting, because accumulation of sugar above a certain concentration temporarily inhibits further production of cellulases, as *adaptive* enzymes. In a study with *Verticillium albo-atrum*, Talboys (1958c) found that production of cellulase in the presence of cellulose was reduced by 0·1% and inhibited by 1% glucose. Similarly, Horton & Keen (1966) reported that production of cellulase by *Pyrenochaeta terrestris* (Hansen) Gorenz, Walker & Larson was repressed by 0·09% glucose.

On present evidence, therefore, we are bound to conclude that production of an antifungal antibiotic is not essential to saprophytic survival by a fungus for its full term in infected or colonized plant tissue, even though it may assist survival if conditions for its production are satisfied. Bruehl, Lai & Huisman (1964) demonstrated production of an antifungal antibiotic by *Cephalosporium gramineum* in 2% cornmeal extract acidified to pH 4·8, and thought that production of this antibiotic helped *C. gramineum* to establish colonies from infected-tissue transplants on the agar medium made from this nutrient solution. Lai & Bruehl (1966) later suggested that continued production of this wide-spectrum antifungal antibiotic by *C. gramineum* in the infected wheat tissue, even at a low concentration, might prolong its saprophytic survival. But a subsequent paper by Bruehl & Lai (1966), on saprophytic survival in colonized wheat straws by four cereal foot-rot fungi and six saprophytic fungi, is entitled 'Prior-colonization as a factor in the saprophytic survival of several fungi in wheat straw'; the conclusion about this situation implied by their title coincides with my own, and with a more general one enshrined in the English proverb 'Possession is nine points of the law'.

SURVIVAL IN TISSUES OF WOODY PERENNIAL HOSTS

A change of scene to problems of saprophytic survival by root-infecting fungi in root systems and collar regions of tree species does not necessitate an immediate change from the main topic of the preceding section, because here again the problem of nitrogen as the main mineral nutrient has attracted recent interest, as exemplified by the work of E. B. Cowling, W. Merrill and their associates in the U.S.A. As Cowling & Merrill (1966) have pointed out, the C/N ratio of wood in different species of tree mostly varies within the range 350–500, though in some species it is as high as 1250. It is apparent, therefore, that wood-inhabiting fungi must exercise a quite remarkable economy in the deployment and conservation of so meagre a supply of nitrogen, to account for the known rapidity with which they can decompose timber. Levi, Merrill & Cowling (1968) have provided evidence for the hypothesis that these fungi conserve nitrogen by recycling it from the older and autolysing regions of the mycelium to the younger and actively metabolizing hyphae. In the presence of glucose, various fungi were found able to utilize several fractions of their own mycelia as sole source of nitrogen in pure culture. Cell-wall nitrogen was taken up less quickly than cytoplasmic nitrogen; experiments both with mycelial fractions and with model nitrogenous compounds showed that proteins, peptides and amino acids were more quickly utilized than nucleic acids, nucleotides or cell-wall fractions.

More recently, Levi & Cowling (1969) have turned their attention to the means whereby a limited supply of nitrogen may be deployed with maximum economy in the mycelia of these wood-destroying fungi. They found that when *Polyporus versicolor* L. ex Fr. was grown on media varying widely in C/N ratio, the nitrogen content of its mycelium ranged from 8·2% on a medium of C/N 4 to one of 0·1% on a medium of C/N 1 600. Comparative analyses of these nitrogen-rich and nitrogen-poor mycelia showed that under conditions of nitrogen starvation, nitrogen was allocated to metabolically active cell constituents such as the nucleic acids, in preference to structural components of the cytoplasm and cell walls. A representative group of white-rot fungi was found able to produce cellulases on a cellulose medium of C/N 2000; other groups of cellulose-decomposing fungi were unable to produce cellulases at this C/N ratio of the culture medium. The white-rot group of fungi are the most highly specialized amongst wood-decomposers, inasmuch as they can degrade lignin as well as cellulose; from these data of Levi & Cowling, it seems that their specialization has extended to include adaptation to the nitrogen-poverty of their substrate.

The massive woody root systems and collar regions of tree species differ so much from the tissues of herbaceous plants as to make the problem of saprophytic survival quite different in most of its aspects. The much greater size of tree roots and stumps, as well as their mechanical construction, greatly reduces the proportion of extraneous nitrogen that can diffuse into the inner tissues from the surrounding soil, at least during the earlier stages of decomposition whilst roots and stumps are still intact. In the absence of wounds, the bark offers a mechanical barrier of cork tissue to fungal invasion; even after this barrier has been penetrated, initial invasion of the wood is limited to cellulose-decomposing fungi. Sylvicultural wounds that expose transverse sections of wood, however, will enable non-cellulose-decomposers, such as the blue-staining species of *Ceratocystis*, to invade the lumina of exposed vessels; another route for more rapid fungal invasion is provided by the tunnels of beetles and other wood-boring insects. Nevertheless, fungal invasion even of dead wood is slower than that of softer plant tissues; mechanical structure, chemical composition (e.g. shortage of nitrogen) and sometimes the presence of fungistatic substances, as in the heartwood of some tree species, all combine to impede the progress of fungal invasion, and to exercise a selective effect upon the species of colonizing fungi.

Most studies so far made of wood colonization by fungi have shown that considerable volumes of wood are often occupied by a single fungal colony. The situation thus often arises in dying or dead tree roots of a meeting between two fungal colonies advancing along the root in opposite directions. This situation has been experimentally studied by Rishbeth (1950, 1951 *a*) in the course of his work on *Fomes annosus*. Root lengths occupied by *F. annosus* were juxtaposed with root lengths occupied by another fungus; competing fungi thus investigated were various isolates of *Trichoderma viride* (*sensu* Bisby, 1939), species of *Ceratocystis* causing blue-stain of timber, and some common basidiomycete competitors of *F. annosus* found in pine stumps, including *Peniophora gigantea* (Fr.) Massee, *Stereum sanguinolentum* (Fr.) Fr. and *Polystictus abietinus* (Dickson) Fr. Amongst the pairings tested by Rishbeth, all three possible results were obtained, i.e. *F. annosus* replacing its opponent or being replaced by it, or a stalemate. Over all pairings, the competitive advantage of *F. annosus* was greater at 10° than at 15° C; this recalls a similar effect of temperature found for several cereal foot-rot fungi and discussed in Ch. 5 (p. 126). Although both protagonists and conditions of competition were quite different in these two situations, they had one thing in common, i.e. that a parasite was competing with other fungi, some of which were obligate saprophytes; this effect of temperature is worth further exploration. Rishbeth also found, as might have been anticipated, that *F. annosus* was displaced less frequently from recently infected roots than from old infected roots in an advanced stage of decay. Lastly, *F. annosus* was found to survive for longer in infected sinker roots lying at some depth in the soil than in the shallower-growing lateral roots, and for longer in alkaline than in acid soils.

Replacement of one fungus by another implies the possibility of a fungal succession in such woody tissues, and such a succession of fungi in unprotected stumps was studied by Meredith (1960) in East Anglican plantations of Scots and Corsican pines. Stump surfaces exposed by tree-felling and left without any of the various protection-treatments now employed (Ch. 8) were at first infected by air-borne spores either of *F. annosus* or of one or other of two common basidiomycete competitors, *Peniophora gigantea* and *Stereum sanguinolentum*. These two latter fungi are weak parasites, inasmuch as they can invade still living stump tissues in which the defence-response to infection has been much weakened by felling of the trunk, but they are not sylvicultural pathogens like *F. annosus*. The basidiomycetes were found to be accompanied in their early invasion of the stump tissues by blue-staining species of *Ceratocystis*, which are non-cellulolytic, and thus occupy an ecological niche distinct from that of the basidiomycetes. These primary basidiomycete invaders were often found to have been replaced in 3 to 4-year-old stumps by *Hypholoma fasciculare* (Hudson) Fr., and after 5 years *Tricholoma rutilans* (Schaeffer) Fr. had become the commonest basidiomycete occupant of these stumps (Table 13).

Table 13. *Succession of basidiomycetes in pine stumps*

(From Meredith, 1960)

| | Percentage stumps occupied by fungi listed, at various ages from tree-felling | | | | | |
| | Age of stumps (years) | | | | | |
	1	2	3	4	5	8
Fomes annosus*	43	37	16	7	17	14
Peniophora gigantea*	85	89	24	3	—	—
Stereum sanguinolentum*	32	29	26	15	6	5
Polystictus abietinus*	—	4	35	13	33	—
Hypholoma fasciculare*	—	—	45	55	24	12
Tricholoma rutilans	—	—	15	43	73	66
Polyporus amorphus	—	—	5	8	4	—
Paxillus atrotomentosus	—	—	—	—	5	4
P. involutus	—	—	—	—	—	1
Collybia maculata	—	—	—	—	4	1
Clitocybe aurantiaca	—	—	—	—	—	3
Flammula sapinea	—	—	—	3	5	2
Pholiota flammans	—	—	—	—	—	1

* Species identified by mycelial characters; remaining species identified by presence of sporophores.

Despite the liability of *F. annosus* eventually to be replaced by other fungi in stump root systems, its longevity in such tissues appears to vary widely according to size of stump and soil type. On an alkaline soil, Rishbeth (1951 *a*) found *F. annosus* to be still alive in 85 % of stumps of 60-year-old pines no less than 25 years after the trees had been felled.

Colonies of wood-rotting basidiomycetes are often bounded in the woody tissue by a distinctive, dark-coloured envelope. In a section across the wood, this envelope is seen as a dark-brown or black, narrow zone, and is therefore called a 'zone-line' because it is viewed in only two out of its three dimensions (and even then this term violates the geometrical definition of a line as 'having length but no breadth'). One of the earliest descriptions of zone-lines was that by Brooks (1915) of the conspicuous black zone-lines produced by *Ustulina zonata* (Lév.) Sacc. in wood of the rubber tree. Campbell (1933, 1934) has described the zone-lines formed by *Xylaria polymorpha* (Pers.) Grev. and *Armillaria mellea*. The boundary envelope of the fungal colony in the wood is formed by a tissue of dark-coloured, sclerotoid and inflated fungal cells. This boundary layer thus resembles the rind of a fungal sclerotium, and so Campbell proposed the name of *pseudosclerotium* for the whole fungal colony in the wood. It cannot be denied that Campbell's analogy is a useful one, provided that the parallel is not pushed too far and allowed to obscure the real physiological difference between these two types of fungal tissue. It is true that, by definition (Ainsworth, 1961), a sclerotium can incorporate host tissue and even soil; nevertheless, the host tissue if present is usually ghost-like in the mature sclerotium, and its nutrients have been transferred to the fungal cells, as in sclerotia of *Helicobasidium purpureum*, formed around small lateral roots on the tuberous tap roots of sugar beet and carrot (Ch. 7, p. 213). So the mature sclerotium of *H. purpureum* and similar fungi represents *dormant survival* of an aggregation of hyphal cells, well stocked with reserves of carbon and other nutrients and comparable to a well filled larder. But in the pseudosclerotium, as defined by Campbell, the fungal larder is still in process of being filled up by active fungal metabolization of the woody host tissue; the pseudosclerotium thus seems to have more in common with saprophytic survival by a colony of still active mycelium than with dormant survival by a mature sclerotium.

To say this, however, is not to declare that the boundary envelope (or zone-line) of the pseudosclerotium lacks the protective function commonly assigned to the rind of a sclerotium; indeed, there is good evidence for the protective function of zone-lines. In a study with several wood-rotting fungi, Hopp (1938) concluded that zone-lines were formed only when the air-moisture régime in wood lay within certain limits; because air and water contents of wood are complementary, as they are in soil, water-saturated wood is oxygen-deficient, and dry wood is water-deficient. Hopp's findings were confirmed by Rishbeth (1951 a), who concluded that gradual desiccation promoted formation of zone-lines by *Fomes annosus* in pine roots. He found that *F. annosus* could be replaced by *Trichoderma viride* (*sensu* Bisby, 1939) and by *Torula ligniperda* (Willkom) Sacc. in root lengths in which zone-lines had either not been formed by *F. annosus* or had been damaged; *F. annosus* was not replaced by these or other fungi when protected by intact zone-lines. Rishbeth has suggested that *F. annosus* surviving in a stump and its roots is protected from replacement by other fungi as a result of zone-lines induced by progressive desiccation, from the stump surface downwards and from the distal region of the roots upwards. Fluctuation in soil moisture content during the summer causes expansion and contraction of the wood, which is likely to result in breaching of the zone-lines, especially if wood decay due to *F. annosus* is in an advanced stage. Continuing desiccation of the decaying wood under a low enough soil moisture content then enables *F. annosus* to form another zone-line nearer to the stump centre, and so the fungus gradually retreats in a succession of rearguard actions, marked by a successive series of zone-lines.

These conclusions about the protective function of zone-lines have been upheld by Nelson's (1964) study of survival by *Poria weirii* in naturally infected wood blocks (2 in³), cut from the heartwood of a Douglas fir, and buried in soil under a young stand of the same species. A sample of thirty-six blocks was taken after

6, 9, 12 and 20 months from burial; at each sampling, presence/ absence of zone-lines in each block was recorded, and viability of *P. weirii* was determined by tissue-plating. The percentage of blocks with zone-lines varied from 69 to 83% over the sampling period. *P. weirii* was recorded from only 10% blocks *without* zone-lines after 6 months in soil, and from none thereafter. But from blocks *with* zone-lines, it was recovered from 100, 90, 70 and 32% at successive sampling periods up to 20 months.

The protective function of the zone-line boundary envelope of the pseudosclerotium was an essential ingredient of the hypothesis proposed by Bliss (1951), to explain the mechanism whereby *Armillaria mellea* in infected citrus roots is killed by soil fumigation with carbon disulphide. As a background to his hypothesis, Bliss noted that *A. mellea* generally persisted for 6 years or more in infected citrus roots of not less than 3 cm diam. in Californian soils; he attributed this considerable longevity to the boundary layer of the pseudosclerotium. In the course of testing for survival of *A. mellea* in infected roots after soil fumigation, Bliss observed that *Trichoderma viride* (*sensu* Bisby, 1939) regularly grew out from tissue platings that failed to yield *A. mellea*, and this led him to suggest that *A. mellea* had been killed not by direct fungicidal action of carbon disulphide, but by antibiotics produced by *T. viride*. Dominance of *T. viride* in soil after fumigation with quite a wide variety of fumigants had already been reported by that time. Bliss therefore suggested that this much augmented soil population of *T. viride* resulting from fumigation was able to penetrate the boundary layer of the *A. mellea* pseudosclerotium, whereas the much smaller population in unfumigated soil was unable to do so. He adduced three kinds of evidence in support of his hypothesis: (1) that the death of *A. mellea* did not immediately follow soil fumigation, at least in the stouter infected roots, but was spread out over a period of some 3 weeks; (2) that *A. mellea* in infected roots could survive a similar fumigation in sterile soil; (3) that *A. mellea* was killed if infected roots were buried in a pure culture of *T. viride*.

It was not incompatible with this very useful hypothesis to admit the possibility that carbon disulphide might kill *A. mellea* within roots of small diameter by direct fungicidal action, and that it might further damage the mycelium of this fungus in the outer part of stout roots within the limits of fumigant-diffusion, and so impair the resistance of *A. mellea* to replacement by *T. viride*. Employing woody inoculum segments of *A. mellea* (approx. 1·75 cm diam.) I found that *A. mellea* could be killed by direct fungicidal action of carbon disulphide in sterile soil, under which conditions no alternative explanation was possible (Garrett, 1957). But I was also able to demonstrate an indirect, microbiological effect of soil fumigation, as postulated by Bliss, upon the viability of *A. mellea*. I buried similar small inoculum segments in soil that had earlier been treated with carbon disulphide at a dosage concentration and period previously ascertained to ensure a maximum development of *T. viride* (*sensu* Bisby, 1939) in the soil after fumigation. After fumigation at this optimum dosage, the soil containers were uncovered to permit the fumigant to disperse from the soil into the atmosphere over a period of 3 weeks; during the third week, green plaques of sporulating *T. viride* could be seen developing between the soil and the glass wall of the containers. At the conclusion of this period, inoculum segments containing *A. mellea* were buried in the fumigated but now fumigant-free soil, and incubated for 23 days; the inocula were then removed and set up in glass tubes filled with soil to assay their capacity for rhizomorph production, as a test for the viability of *A. mellea* within each inoculum segment. In a comparison between twenty inoculum segments buried in unfumigated soil and twenty buried in soil fumigated with an optimum dosage of carbon disulphide, I found that thirty-nine of forty segment-ends from the control series produced rhizomorphs within 26 days in the growth tubes; in the series from the fumigated soil, only twenty-eight of forty segment-ends had produced rhizomorphs after 75 days. Incubation of the inoculum segments for 23 days in the fumigated but fumigant-free soil had thus caused 30% of segment-ends to lose their ability to produce rhizomorphs. This effect must have been due to some change in the soil brought about by fumigation; taken in conjunction with the visible development of *T. viride* at the soil–glass interfaces of the containers, it seemed to support at least some of the claims made by Bliss.

Bliss had also reported a complete kill of *A. mellea* in infected roots buried in a pure culture of *T. viride*, and this suggested exploration of the possibility that the inoculum potential of *T. viride* was a determinant of success in its lethal action against *A. mellea*, as was indeed explicitly postulated by Bliss. In an experiment to test this (Garrett, 1958), I prepared pure-culture inoculum of *T. viride* on autoclaved soil, and aliquots of this were diluted to one-quarter and one-eighth strengths with the same soil unsterilized. Twenty inoculum segments were buried in each of these soil mixtures for 23 days, and all segments were then transferred to soil growth tubes to assay capacity of inocula to produce rhizomorphs, as before. *A. mellea* failed to produce rhizomorphs, and was therefore presumed dead, in 95% of inoculum segments buried in the pure culture of *T. viride*. Amongst the twenty inoculum segments buried in one-quarter strength inoculum of *T. viride*, 25% of segment-ends failed to produce rhizomorphs within 76 days; this loss of rhizomorph-producing capacity was thus closely comparable to that following incubation of inoculum segments in a soil that had received a fumigant dosage optimum for subsequent development of *T. viride* in the soil. Taking Bliss's results and my own together, I concluded that some post-fumigation biological control of *A. mellea* by *T. viride* could occur under optimum conditions, though less than Bliss had originally claimed. More modestly on behalf of *T. viride*, I preferred to say that *T. viride* could help to finish off what fumigation had begun. At that time, I did not appreciate that populations of *T. viride* (*sensu* Bisby, 1939) in fumigated soils could vary very widely in their antagonistic effects against *A. mellea* and other fungi, as will shortly transpire. Both Bliss and I were lucky in obtaining positive results, though whether it is 'lucky' to obtain results that can be confirmed only with difficulty by subsequent workers is open to question.

Reasons for the development of dominance by *Trichoderma viride* (*sensu* Bisby, 1939) after soil fumigation with a suitable dosage of carbon disulphide have been explored by Saksena (1960) and Moubasher (1963). Both these workers found in fumigation trials with a range of carbon disulphide doses that *T. viride* possesses only a moderate degree of tolerance to carbon disulphide; various species of *Penicillium* and *Aspergillus*, mostly ascosporic, can survive a dose at least five times that tolerated by *T. viride*. Saksena therefore concluded that *T. viride* owes its ability to dominate the fungal population recolonizing fumigated soil to possession of a moderate but sufficient degree of fumigant tolerance allied with a high growth rate through fumigated soil; some fungi with a still higher growth rate possess only a low fumigant tolerance whereas the highly tolerant fungi are mostly rather slow growers. This combination of these two characteristics thus possessed by *T. viride* appeared to give it a unique advantage for recolonization of fumigated soil, at least amongst the common soil fungi studied by Saksena and Moubasher. The fuller data collected by Moubasher are compatible with Saksena's explanation, and he also provided a useful quantitative expression for *recolonizing ability*, in the ratio obtained by dividing the soil population of a particular fungal species 36 days after fumigation by its population 1 day after fumigation. He estimated soil populations by soil-dilution-plate counts. This recolonizing ratio for *T. viride* was 9·2, and was higher than that for any of the other twenty-three common soil fungi that Moubasher investigated.

Analysis of this problem was carried further by Mughogho (1968), who began work in the hope that he could find a fumigant more reliable for promoting dominance of *T. viride* than we had found carbon disulphide to be. To obtain even a moderate development (as compared with that in sterile soil) of *T. viride* in fumigated soil, the dosage of carbon disulphide had to be kept within a narrow range and the prospects of being able to do this by soil fumigation in the field were not encouraging. Moreover, in my own fumigation experiments with carbon disulphide, the indirect component of the fumigation effect on *A. mellea*, although significant, had not been good enough for a control measure in actual practice, and we hoped to improve on this by finding another fumigant more effective for this particular purpose. Mughogho's first results in this search were encouraging, as the following fumigants were found superior to carbon disulphide for promoting dominance of *T. viride*, both in respect of breadth of suitable dosage range, and in the size of *T. viride* populations developing in the fumigated soils: allyl alcohol, chloropicrin, methyl bromide, D-D mixture, sorbic acid and acetylenedicarboxylic acid. It was therefore unexpectedly disappointing for Mughogho to find that these high populations of *T. viride* in the fumigated soils more often than not failed to exert any controlling effect either on production of rhizomorphs by *Armillaria mellea* or on mycelial growth through soil and infection of swede seedlings by *Rhizoctonia solani*. Most fortunately, however, it became possible to explain these unexpected results through the current research by J. Webster and M. A. Rifai at the University of Sheffield on a revision of the taxonomy of the genus *Trichoderma* (Webster, 1964; Webster & Lomas, 1964; Rifai & Webster, 1966*a*, *b*; Webster & Rifai, 1968; Rifai, 1969). In 1939, G. R. Bisby revised the taxonomy of the genus *Trichoderma* Pers. ex Fr. and incorporated all described species into *Trichoderma viride* Pers. ex Fr., thus making the genus a monotypic one. But Webster and Rifai have now found good taxonomic reasons for putting asunder what Bisby had joined together, and Rifai (1969) has published a conspectus of the genus *Trichoderma* in which nine species are recognized and keyed for identification.

The species, as now recognized by Rifai (1969), that Mughogho found to make up his populations of *Trichoderma* in these variously treated soils were *T. hamatum* (Bon.) Bain., *T. harzianum* Rifai, *T. koningii* Oud., and *T. viride* Pers. ex S. F. Gray; amongst these four species, *T. harzianum* usually predominated. He further found that different isolates of *Trichoderma*, even within a single species, varied widely in their degree of antagonism towards *Armillaria mellea* and *Rhizoctonia solani*. He concluded that dominant populations of *Trichoderma* in his fumigated soils must have been made up largely of isolates that were not effectively antagonistic towards either *A. mellea* or *R. solani*. The highest proportion of effective antagonists amongst isolates of *Trichoderma* tested by Mughogho belonged to *T. viride* Pers. ex S. F. Gray, as now delimited by Rifai (1969).

It cannot be denied that all these studies of the taxonomy and ecology of *Trichoderma* species in soil have greatly increased our comprehension of the sequence of mycological events occurring after soil fumigation; equally, however, we have to admit that our hopes of mediating biological control of root-disease pathogens by *Trichoderma* species through soil fumigation, originally aroused by D. E. Bliss in 1951, have received a set-back. My account of these events, however, is merely a progress report; it is not an epitaph.

7 Dormant survival by resting propagules of root-infecting fungi

Dormant propagules constitute the most effective mode of survival by root-infecting fungi between one host plant and the next; infective propagules with a period of dormancy are of the following kinds: (1) *dispersal spores,* which are produced either sexually or asexually by the great majority of fungi in all major taxonomic groups; (2) *resting spores* resulting from sexual reproduction, such as the oospores of the Oomycetes and asco-spores of some species of the Ascomycetes; (3) thick-walled *chlamydospores* formed within hyphal and conidial cells, especially by species of *Fusarium*; (4) *sclerotia*, which are multicellular resting bodies produced by fungal species in all taxonomic groups except the Phycomycetes. In a review of fungal spore germination, Gottlieb (1950) produced the following definition of *dormancy*: 'A spore is dormant when it does not germinate under the same nutritive and environmental influences which later allow germination'. After morphological demarcation of the spore, a further period of maturation must ensue before it is capable of germination under suitable environmental conditions. As Gottlieb has pointed out, a period of maturation is essential for spores of all types, and it is only the length of this period that distinguishes dispersal spores from resting spores. In some resting spores, such as the oospores of *Phytophthora cactorum* (Leb. & Cohn) Schröt., this maturation period may be both prolonged and of very variable duration in different individuals (Blackwell, 1943 *a*, *b*).

The subject of spore germination has more recently been reviewed by Gottlieb (1966), Sussman (1966) and Cochrane (1966). Dormancy as defined by Gottlieb (1950) has been qualified by Sussman as *constitutive dormancy*, to distinguish it from *exogenous dormancy* imposed upon the spore by environmental conditions that inhibit germination. As an example of constitutive dormancy, Sussman has cited the ascospores of *Neurospora* species, which require a heat-shock before they will germinate during the dormant period. Constitutively dormant spores often or even usually can germinate when ripe without an exogenous energy source, whereas for other spores this is often an essential. Cochrane (1966) has noted that the relatively short-lived asexual spores of the Ascomycetes and Fungi Imperfecti usually require exogenous sources of both carbon and nitrogen nutrients. He has termed such spores *heterophagous*, in contrast to *autophagous* spores that will germinate without exogenous nutrients. Cochrane has also emphasized the importance of certain precautions when experimentally determining a possible need for exogenous nutrients in spore germination. Washing spores off a nutrient agar surface will dissolve some of the substrate nutrients; if then spores are centrifuge-washed to get rid of these, endogenous nutrients may be reduced by leaching below the critical level for germination. It is therefore best to remove spores from the parent culture with a vacuum collector, or by brushing, shaking, etc.; the choice of method will depend upon whether the fungal sporophores produce dry spores (xerospores) or slime spores (gloiospores).

SPORE GERMINATION IN THE SOIL

An exogenous dormancy is imposed upon fungal spores lying in the soil by a fungistatic factor, or factors, that is widespread in most soils, both natural and cultivated. General interest in this problem of soil fungistasis was first aroused by Dobbs & Hinson (1953), who reported that spores of various soil fungi that germinated freely in distilled water would not do so in soil solution diffusing through a cellophane membrane. Simmonds, Sallans & Ledingham (1950) had earlier observed a similar effect on spore germination, but with the conidia of just a single fungus, *Cochliobolus sativus*, so that their report failed to arouse the widespread interest that greeted the announcement by Dobbs & Hinson, who also demonstrated that a glucose solution of concentration only $0 \cdot 1 \%$ was sufficient to override the fungistasis and cause spores to germinate. In spite of all the work that has been reported since that time, the overcoming of soil fungistasis by glucose and other nutrients can still be best interpreted through the profoundly important generalization made by William Brown (1922*b*) as a result of his study of the effect of carbon dioxide upon spore germination and mycelial growth of various mould fungi; Brown epitomized his findings by saying that the inhibiting effect of carbon dioxide was greatest where the fungal energy of growth was least. These observations by Dobbs & Hinson were soon confirmed by those of Jackson (1958*a*, *b*) for a number of Nigerian and English soils, and with some other species of soil fungi, some of which were unaffected by the general soil fungistasis. In two soil profiles studied, fungistatic effects were confined to the upper 40 cm. The soil fungistatic effect was found to decrease with increasing acidity, suggesting to Jackson that soil bacteria might be important as producing agents. It is important to remember, however, that a sufficient degree of acidity in the soil solution can of itself reduce the percentage of spores germinating in any population. Dobbs & Gash (1965) have opportunely distinguished between *microbial fungistasis,* as originally reported by Dobbs & Hinson, and *residual fungistasis.* Microbial fungistasis is considered to result from the products of microbial metabolism, because it can be removed by autoclaving and by all sterilizing agents studied, and does not return so long as a sterilized soil remains sterile; it is usually absent, or virtually so, from subsoils, which have a low microbial population. Residual fungistasis, on the other hand, was found by Dobbs & Gash to be thermostable, and was not removed by soil sterilization. I have already suggested that a residual fungistasis of this type could be caused by a sufficiently extreme degree of soil acidity, but that studied by Dobbs & Gash was observed in calcareous soils. They detected this residual fungistasis in sand from the foreshore and mobile dunes of the Anglesey coast, but not in that from the fixed dunes, with a higher organic content. Their experiments eliminated calcium carbonate as a direct cause of residual fungistasis but various results agreed in suggesting that the toxicity was due to unchelated iron at the high pH value (8·0 or above) of the sand; chelation of the iron by the organic matter that had accumulated in the fixed dunes could thus provide an explanation for the disappearance of this residual fungistasis with increasing age of the dunes.

Much has been contributed towards solution of this problem of microbial fungistasis by the intensive studies of J. L. Lockwood and his associates in the U.S.A. Lingappa & Lockwood (1961) failed to extract any *persistent* soil toxin and therefore suggested that a fungistatic level must be maintained by continuous microbial production, especially by spore-surface bacteria subsisting on spore-exudate nutrients. The same authors later (1964) reported a several-fold increase both in bacterial numbers and in soil respiration when soil was supplemented either with living fungal spores or with aqueous washings of the same. When washed cells of bacteria or of *Streptomyces* species were incubated with fungal spores in absence of any exogenous nutrients, spore germination was thereby inhibited. In a contemporaneous review, Lockwood (1964) noted that *washed* fungal spores germinated poorly or not at all in distilled water and suggested that bacteria and actinomycetes might prevent fungal spore germination in soil by consuming spore-exudate nutrients as quickly as they were released through the spore membrane. This hypothesis clearly implies that the germination of similar but *unwashed* spores in distilled water is due to a back reaction of the spore exudate upon the spore that has produced it; when the exudate has attained a certain threshold concentration in the water immediately surrounding the spore, it must trigger off germination, according to this hypothesis. As I have mentioned already, and as we shall see in more detail later, many types of spore require exogenous carbon nutrients and some need exogenous nitrogen as well before they will germinate. Such observations constitute part of the evidence in support of Lockwood's hypothesis; successful spore germination in the absence of exogenous nutrients thus seems likely to depend not only on the total content of endogenous nutrient reserves, but also on their rate of mobilization in soluble form to the critical threshold concentration.

Continuing to emphasize this nutrient-deficiency component of microbial soil fungistasis, Chacko & Lockwood (1966) criticized the cellophane film employed by Dobbs & Hinson (1953) and the water-agar discs used by Jackson (1958*a, b*) as spore carriers, on the grounds that neither of these carriers was completely nutrient-free. They recommended instead a more direct method devised by Lingappa & Lockwood (1963) whereby the fungal spores were set to germinate on a thoroughly smoothed surface of finely sieved soil, and after an incubation period were stained and then fixed by impregnating a very thin surface layer of the soil with collodion; this was then peeled off when dry as a thin film for microscopical examination. Employing this technique, Chacko & Lockwood assessed the strength of soil fungistasis by incubating spores on the surface of a graded dilution series of a natural soil with the same soil sterilized by ethylene oxide vapour. Increasing dilution of the natural with the sterilized soil caused two progressive changes: (1) by provision of an increasing amount of nutrient substrate in the form of sterilized soil; (2) by a concomitant reduction in the density of the microbial population in the mixture; each of these effects must have contributed to a gradual decline in the intensity of fungistasis. This technique thus provided a method for quantitatively comparing both different soils for intensity of the fungistatic effect, and spores of different species of soil fungi for their susceptibility to this effect, through the ED50 value, i.e. the proportion of sterile soil in the mixture just sufficient to permit 50% spore germination. The ED50 varied for different species of fungi in the same soil, as earlier work had shown. Intensity of fungistasis was found to decrease rapidly down the soil profile, from a value of 0·7 at 10 cm down to one of 0·05 at 61 cm. In a trial with eighteen species of fungi, Ko & Lockwood (1967) correlated ability of spores to germinate in natural soil with their ability to germinate in absence of exogenous nutrients ($r = 0.94$; $P < 0.5\%$). But the spores of four species that were nutritionally independent in germination on glass slides were found not to germinate in the soil; this was ascribed by Ko & Lockwood to a steep concentration gradient of spore-exudate nutrients from spore surface to soil caused by microbial assimilation at the spore surface, as originally suggested by Lockwood (1964). When spores are standing in a drop of sterile water on a glass slide, the nutrient concentration gradient due to diffusion is much more shallow. But when Ko & Lockwood increased the

12-2

steepness of this concentration gradient in another way, by placing spores on filter paper continuously leached with distilled water, germination failed to occur. When they provided a substrate by incorporating ground dried alfalfa (*Medicago sativa*) tissues with the soil, then all spores, including those of nutrient-dependent species, germinated. Chinn (1953) had earlier reported a similar effect of an amendment of soybean meal in inducing spore germination of *Cochliobolus sativus*, *Fusarium roseum* f. *cerealis* and seven other species in soil.

The use of water-agar discs as a carrier for the spores to be tested againt soil fungistasis by Jackson (1958*a*, *b*) has already been noted; in this technique, fungistatic substances that are water-soluble diffuse through the agar, as they do through the cellophane film employed by Dobbs & Hinson (1953). Dix (1967) improved upon Jackson's technique by slicing agar blocks into a graded series of thicknesses (200, 400, 600 and 800 μ) with a microtome, so that the concentration of water-soluble fungistatic substances was graded by diffusion according to the thickness of the agar. His addition of 1% peptone to the agar is likely further to have widened the spectrum of germination-response amongst the twelve species he studied, which were chosen from amongst a number isolated from the root surface and rhizosphere of *Phaseolus vulgaris* (Dix, 1964). A wide spread in the effect of soil fungistasis upon spore germination was found amongst these twelve species; *Penicillium janthinellum* Biourge and *Trichoderma koningii*, which were pioneer root-surface colonizers (apical 1 cm of roots) were amongst the most tolerant. In a comparable study at Cambridge, we have confirmed Dix's findings on the tolerance of these two species to fungistatic effects, and also his observation that *Mucor plumbeus* Bon. and *Penicillium nigricans* (Bain.) Thom are rather sensitive to fungistasis (Dwivedi & Garrett, 1968). Dix's experiments were carried out with washed spores and he was thus able to determine that three of his species, *Penicillium nigricans*, *Mucor plumbeus* and *Doratomyces purpureofuscus* (Fr.) Morton & Smith, required an exogenous supply of both carbon and nitrogen nutrients for spore germination; *M. plumbeus* further required one or more vitamins present in yeast extract.

The agar spore-carriers employed both by Jackson and by Dix make it probable that the fungistatic factors they studied were water-soluble but not necessarily volatile. Balis & Kouyeas (1968) have made use of the ability of silver and mercury salts to form complexes with volatile hydrocarbons to demonstrate the occurrence of a volatile fungistatic factor, possibly of this type, in the soils they studied. The fungistatic property was lost when soil was air-dried, but was restored when the air-dried soil was remoistened and incubated; restoration was most rapid and complete at a moisture content optimum for microbial activity, and did not occur in water-saturated soils (Kouyeas & Balis, 1968). Fungistasis was annulled by passing an airstream over the sporiferous soil surface. Disappearance of fungistasis on air-drying of soil was ascribed to loss of the adsorbtive properties of clay mineral particles when air-dried.

We can now attempt to put this recent work on spore germination in the soil into perspective against the background of earlier work on nutritional factors affecting spore germination, which has been reviewed by Gottlieb (1950) and Cochrane (1958). It now seems likely that most thin-walled dispersal spores are nutrient-dependent in germination, at least partially. Some are absolutely dependent upon one or more exogenous nutrients; most others are partially nutrient-dependent, in the sense that percentage germination is usually higher, and often much higher, in a nutrient solution than in pure water. The greatest response is usually to sugars. The laboratory situation in which spores are set to germinate on a glass slide is not so far removed from that in a film of water overlying a leaf; as Brown (1922a) showed with spores of *Botrytis cinerea*, germination-promoting substances diffuse through the cuticle, so that percentage germination is likely to be higher than that in water on a glass slide, though lower than that in a nutrient solution of optimum composition and concentration. But the situation of a spore germinating on a root surface is further removed from that on a glass slide; the more rapid exudation of nutrients from the root surface as compared with that through the cuticle, the often higher relative humidity of the soil atmosphere and the absence of extreme temperature fluctuations all combine to promote a degree of microbial activity at the root surface that is much higher than that on the leaf surface in most types of climate.

We can now understand the biological implications of this probably fairly general difference in behaviour between spores of leaf-infecting and root-infecting fungi, respectively. As Kerr & Flentje (1957) were the first to point out, the spore of a leaf-infecting fungus once deposited on a leaf and stuck there has only two alternatives: to infect or to perish. There is thus no visible advantage in exogenous dormancy imposed by a critical need for exogenous nutrients in germination. But with spores of root-infecting fungi, nutrient-imposed exogenous dormancy prevents suicidal spontaneous germination in the absence of a substrate stimulus; because roots grow through the soil, waiting spores are almost certain to come into contact with a substrate sooner or later.

The behaviour of thick-walled resting spores is determined by a more complex set of variables. Some of them can certainly germinate quite well in pure water, once they are ripe to do so. We do not know how important dormancy-breaking treatments may be in the natural environment; in the laboratory they provide a convenient means of saving time and imposing a greater uniformity in germination on a spore population. Such treatments as heat-shock, exposure to low temperature and to certain chemicals may affect either the permeability of the spore wall or the state of the protoplast, or both. Permeability is not necessarily related to wall thickness; a thin wall may be as impermeable as a thick one. Germination-stimulating chemicals that are not nutrients and act at a low concentration raise further problems. Various volatile substances are effective activators of spore germination, and are responsible for the germination-promoting effect of some fresh plant tissues, as first demonstrated by William Brown (1922c). Noble (1924) found that suitably low concentrations of some volatile chemicals, such as benzaldehyde, salicylaldehyde and acetone, stimulated germination of the chlamydospores of *Urocystis tritici* Körn., causing flag smut of wheat. Hooker, Walker & Link (1945) found a similar effect of a mustard oil, allyl isothiocyanate, on germination of the resting spores of *Plasmodiophora brassicae*, and suggested that this and similar volatile substances might be responsible for the germination-promoting effect of cruciferous host roots. King & Coley-Smith (1968) have reported that a volatile germination-activator for the sclerotia of *Sclerotium cepivorum* Berk. is given off by growing seedlings of onion, leek and garlic; also stimulatory was the pure chemical diallyl sulphide.

A valuable insight into the complex of factors that determine the nature and duration of constitutive dormancy in thick-walled resting spores was early provided by Blackwell's (1943*a*, *b*) intensive study of the oospores of *Phytophthora cactorum*; she has described their development, dormancy and eventual preparation for germination as follows. After fertilization by the antheridium, the young binucleate oospore, lying loosely within the oogonial wall, forms an *exospore* upon the inside of the oospore membrane. This exospore, composed of pectic substances and proteins, is thin, smooth and transparent; it seems to be virtually impermeable, as few stains were found able to penetrate it. Within the exospore is then laid down the *endospore*, which eventually attains a thickness of $2\,\mu$; it is composed of hemicellulose, cellulose and protein and seems to function as a food reserve, because it is later eroded during preparation for germination.

Three stages have been distinguished by Blackwell in the development of this oospore:

(1) cytological development; the period from fertilization to fusion of the two nuclei takes about 1 month at laboratory temperature (15–$20°$ C), but longer at lower temperatures;

(2) after-ripening period; this occupies 6–8 months at 15–$20°$ C, but only 1–2 months at a temperature just above $0°$ C.

(3) preparation for germination; nuclear changes are accompanied by digestion of the endospore, and by desaturation and assimilation of the oil globule, and these events take at least a month (Fig. 20).

The most important conclusion drawn by Blackwell from this thorough and prolonged study was that individual oospores varied widely in their period of dormancy and in their germination-response to environmental conditions. As she has stressed, this wide physiological heterogeneity amongst a population of oospores promotes maximum longevity of the population, because under no

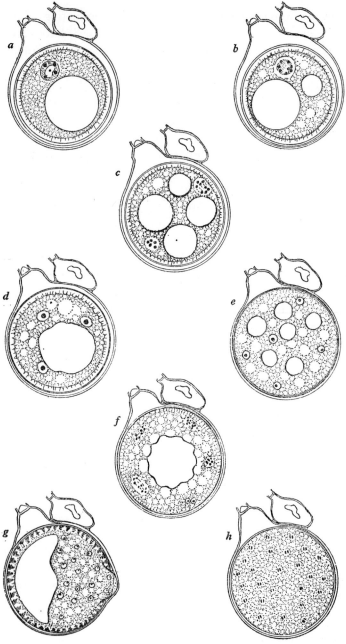

See facing page for legend

single set of environmental conditions will more than a minor proportion of the population germinate at any one time. When hosts and substrates are absent, summer may be a more hazardous season for survival of the species than is winter, but the oospore population provides against all environmental contingencies.

SURVIVAL OF SPORES

In this section I shall give an account of the behaviour in soil of various types of spore, selected as a representative range from amongst the most studied root-infecting fungi, as follows: asco-spores of *Ophiobolus graminis*, basidiospores of *Fomes annosus*, conidia of *Cochliobolus sativus*, macroconidia and chlamydospores of various *Fusarium* species, ascospores of *Rhizina undulata* Fr. and resting spores of *Plasmodiophora brassicae* and *Spongospora subterranea* (Wallr.) Lagerh. Upon all these spores when lying in soil, the general soil fungistasis imposes an exogenous dormancy, which can be broken by a suitable host or substrate stimulus; all of them can thus be said to function as dormant propagules, even though the total life of the spore population in soil may be short, as in the ascospores of *Ophiobolus graminis*, which are primarily dispersal spores of delicate structure and ephemeral existence (Samuel & Garrett, 1933). At the other end of the scale come the ascospores of *Rhizina undulata*, which are characterized by a well marked constitutive dormancy that can be terminated precociously by heat-shock. Similarly the resting spores of the Plasmodio-

20. Oospore of *Phytophthora cactorum* towards the end of dormancy, between 'after-ripening' and germination: (*a*) and (*b*) fusion nucleus in division, oil globule breaking up, endospore becoming eroded; (*c*) second nuclear division, oil globule broken up; (*d*) four-nucleate stage, oil globule dispersing gradually; (*e*) eight-nucleate stage, oil globule further subdivided, endospore almost completely absorbed; (*f*) fourth nuclear division, oil globule still single, endospore almost completely absorbed; (*g*) about thirty-two nuclei, oil globule still single, endospore not yet absorbed, germ tube through exospore but not yet through oogonium wall; (*h*) sixth nuclear division, nuclei very small, all reserves absorbed into protoplasm but germ tube not yet formed. (*a*)–(*h*) are composite drawings of median longitudinal sections (× 1000) not representing a sequence of stages. (After Elizabeth Blackwell, 1943*a*.)

phorales pass through a period of constitutive dormancy, though this appears to be of brief duration in some 95 % or more of the population; it is the longevity of the remaining 5 % or less of the resting spore population that constitutes the problem for plant pathologists.

Ascospores of Ophiobolus graminis

Whether these ascospores could function as infective propagules remained an unsolved problem for many years. In many inoculations of the soil around wheat seedling roots, I failed completely to secure infection, though it was easy to infect roots of seedlings growing in completely sterile soil or sand (Garrett, 1939). I therefore concluded that these ascospores were so completely dependent upon exogenous nutrients in the root exudate that they could cause infection only when the exudate was wholly available to them, i.e. in the absence of other soil micro-organisms as in sterile sand. The corollary that the ascospores were most unlikely to infect roots growing in natural soil was later questioned by the occurrence of widespread outbreaks of take-all amongst early wheat crops on the recently drained Noord Oost Polder of the former Zuider Zee in the Netherlands, which I was shown round in 1946 (Bosma, 1946). The soil of this Polder was virgin in the most literal sense of this word, because it had been drained from under the fresh water of the Ysselmeer, and so *Ophiobolus graminis* must have been transported thither from outside the Polder. Several attempts to secure infection of roots of susceptible grasses by soil inoculation with ascospores ended equally in failure, and the problem was finally solved by D. H. Brooks (1965). He demonstrated that ascospores could readily infect the proximal part of the seminal roots of wheat seedlings that had been produced by allowing wheat seeds to germinate *on the surface* of wet soil. This proximal part of the seminal roots, between the seed and the soil, is not in direct contact with the soil and so is at first virtually free of a root-surface microflora. If ascospores are inoculated (or naturally alight) upon this infection site, then seed and root exudates are at first available to them in virtual absence of microbial competition.

Compatible with this interpretation was Brooks's finding that if seedling roots were left for a period after emergence before they were inoculated with ascospores, then the longer the delay the greater was the reduction in proportion of successful inoculations.

We concluded from Brooks's discovery that ascospore infection by *O. graminis* could occur in the field when wheat seeds were germinating, at a time of ascospore dispersal, on the surface of sufficiently moist soil; these conditions are fulfilled by rainy weather around harvest time, when some wheat seeds are shed from overripe ears, as witnessed by the not infrequent occurrence of 'volunteer' wheat plants in a following crop of another species. We further appreciated the possibility that the low microbial population of the virgin soils of the Netherlands polders immediately after draining might be insufficient to prevent ascospore infection of wheat or susceptible grass roots *within the soil*. Since then, this possibility has been confirmed in some most ingenious experiments by Gerlagh (1968) on the East Flevoland Polder, drained in 1957. For these experiments, he grew wheat seedlings in pots of subsoil clay taken from under the floor of the Polder, with due precautions to avoid contamination by the air spora; in this way, he obtained a 'soil' closely comparable to that emerging when the water had first been drained off the Polder. He soon noted that spontaneous infection by *O. graminis* frequently developed on root systems of wheat plants growing in pots of subsoil clay that had *not* been inoculated with the pathogen, and suspected that these infections had arisen from ascospores present in the air spora of his glasshouse. He followed this up by suspending perithecia-bearing wheat stubble over plants of wheat, *Lolium perenne* and *Poa annua* growing in pots of subsoil clay. All three species became infected by the ascospores. The results of these glasshouse experiments were confirmed by a small field trial, in which a plot was excavated to a depth of 1 m, lined with plastic sheeting and filled up with fresh subsoil clay. The second season's wheat crop on this plot developed numerous whiteheads due to *O. graminis*. In conjunction with these experiments, Gerlagh made a microscopical examination of ascospore-inoculated roots of wheat seedlings growing in Perspex observation boxes (Gerlagh,

1966). In the subsoil clay, roots were regularly infected; in an old cultivated soil, only few plants became infected and lysis of ascospore germ tubes was observed.

Gerlagh has concluded that *O. graminis* first establishes itself by ascospore infection of the roots of susceptible grass species, which are amongst the pioneer colonizers of the freshly exposed floor of a polder after draining. He verified this supposition by finding widespread infection of grass root systems all over an artificial island in one of the polder lakes; this island was made by pumping bottom sludge into a walled enclosure, and its mode of construction eliminated the possibility of contamination by any infective propagules of *O. graminis* other than the ascospores.

Basidiospores of Fomes annosus

In the course of his early work that demonstrated widespread infection of coniferous stump surfaces by the air-borne basidiospores of *F. annosus*, Rishbeth (1951 a) incidentally discovered that these dispersal spores can survive as dormant propagules in the soil, a discovery that certainly surprised me at that time. This finding arose from a trial of covering freshly exposed stump surfaces with soil as a possible protection against basidiospore infection, as had been recommended against other air-borne fungal pathogens in the tropics; in this trial, some 50% of the soil-covered stumps became infected by *F. annosus*, in the absence of any natural inoculum other than basidiospores, and this actually exceeded the percentage infection of the untreated control stumps. Rishbeth verified this conclusion by an experiment planned for the purpose, and also discovered that soil from under a coniferous plantation was infective to a depth of more than 50 cm, which was explicable only on the assumption that basidiospores get washed down through the soil during and after periods of heavy rainfall. This again is an example of exogenous dormancy, presumably imposed by soil fungistasis.

Conidia of Cochliobolus sativus

Studies of these conidia in Canada have been particularly concerned with their exogenous dormancy in soil and how far this varies according to the particular isolate of the fungus. When conidia of a fungus, whether lying in soil or after introduction into it, germinate in the absence of any discernible stimulus from a host root or from any other substrate, their behaviour is described as *spontaneous germination*. Chinn & Tinline (1963) have coined the phrase 'inherent germinability' to express the tendency towards spontaneous germination amongst isolates of *C. sativus*; the degree of inherent germinability is to be equated with the percentage of spores spontaneously germinating. To determine the range of inherent germinability amongst a collection of 255 wild-type isolates, glass slides coated with a suspension of conidia in water agar were set in the soil, and the percentage of spores subsequently germinating was then recorded. Spontaneous germination amongst these isolates ranged from 0 to 87 %, though the majority (135) showed none at all. When reviewing this subject earlier (Garrett, 1956 a), I suggested that spontaneous germination was a departure from our conception of ideal behaviour amongst a population of resting propagules, and must result in considerable wastage. This speculation now rests on the firmer ground of evidence provided by Chinn & Tinline (1964), who estimated respective longevities in soil of spore populations varying in inherent germinability from 0 to 65 %; longevity was estimated in two ways: (1) by recovering spores from soil by the oil-flotation method of Ledingham & Chinn (1955); (2) by a wheat seedling test for viability of spores *in situ* in the soil. An inverse correlation was found between inherent germinability of an isolate and longevity of its spore populations in soil, and so expectations were duly confirmed. Continuing these studies, Chinn (1967) devised a dilution technique to provide a more sensitive assay for the degree of fungistasis in different soils, and for the sensitivity to fungistasis of various isolates of *C. sativus*. The method was similar to that of Chacko & Lockwood already described, except that dilution of the natural soil was made with inert silica particles (grade 70 mesh/in) instead of with the same soil sterilized. Amongst eleven soils tested, strength of fungistasis was greatest in peat soil, intermediate in clays and loams, and least in sandy soils. With the most sensitive isolate of *C. sativus*, spore germination was completely inhibited by addition of only 0·25 % of a loam, clay or peat soil to the silica medium, or by addition of 1 % of a sandy soil.

Up to the time of the most recent of the studies cited above, dormancy of these conidia of *C. sativus* was assumed to be maintained by soil fungistasis. There is still no need to challenge this assumption, but the recent announcement by Meronuck & Pepper (1968) of their discovery of chlamydospores within conidial cells of *C. sativus* does make it easier to comprehend the considerable longevity of some of these spore populations. Meronuck & Pepper introduced conidia into pots of autoclaved soil, which were then sunk in soil outside and left to experience the North Dakota winter. About 70 % of the 800 conidia recovered from the potted soil by the oil flotation technique, during and after the winter, showed internal chlamydospores as evidenced by thickening of the inner walls of one or more cells.

Macroconidia and chlamydospores of Fusarium *species*

Early observations on the behaviour of macroconidia of *Fusarium roseum* f. *cerealis* in soil were made by Chinn (1953) in parallel with his study of the conidia of *Cochliobolus sativus*, just noted. When 2 % soybean meal was added to the soil, the macroconidia of *F. roseum* germinated and produced a limited mycelium on which chlamydospores were formed, thus ensuring continued survival of the fungus at an augmented population level. These observations were the precursor of many more.

The need of macroconidia of *Fusarium* species for exogenous nutrients in germination has been established for several species and may be fairly general. Sisler & Cox (1954) reported that washed macroconidia of *F. roseum* required both exogenous carbon and exogenous nitrogen, and a higher percentage of germination was obtained only in presence of some other constituents of the basal nutrient medium as well; similar results were obtained by

Marchant & White (1966). Cochrane, Cochrane, Simon & Spaeth (1963) found that macroconidia of *F. solani* f. *phaseoli* required exogenous carbon and nitrogen and a growth factor present in yeast extract, which latter could be replaced completely by ethanol or acetoin, and partially by acetaldehyde or one of several amino acids. Rather surprisingly, perhaps, the requirement for an exogenous carbon nutrient was found to exceed the total dry weight of the spore, suggesting that endogenous reserves were inadequate for germination and that substrate supplementation, in addition to germination-activation, was an essential requirement supplied by exogenous carbon nutrients. This suggestion was reinforced by the finding that the requirement for exogenous carbon nutrients and ethanol could be reduced, though not eliminated, by producing the macroconidia on a high carbohydrate medium.

Nutrient requirements for infection are higher than those for spore germination and so the nutrients supplied by root exudate must generally be essential for infection. Toussoun, Nash & Snyder (1960), investigating infection of hypocotyls of snap bean (*Phaseolus vulgaris*) by macroconidia of *F. solani* f. *phaseoli*, determined that exogenous glucose was essential for the prepenetration stage of infection, i.e. germination of conidia and epiphytic development of mycelium on the hypocotyl surface; nevertheless, glucose without nitrogen delayed host penetration and the onset of pathogenesis. Nitrogenous nutrients, on the other hand, favoured early penetration and pathogenesis; organic forms of nitrogen were more effective than inorganic ones. This line of investigation has been further developed, but with chlamydospores instead of macroconidia, and so will come up for discussion again later.

In discussing the formae speciales of *F. oxysporum* causing vascular wilts (Ch. 3), I noted the growing consensus of opinion that chlamydospores are the propagules responsible for the great longevity of some of these pathogens in soil free from host plants. According to Snyder & Toussoun (1965), chlamydospores are produced by most clones of *F. solani*, *F. oxysporum* and *F. roseum*, but not commonly by the other six species of this genus. Snyder &

Hansen (1945) recognized nine species, which they distinguished according to shape of the macroconidia, presence/absence of microconidia and their position and kind, and presence/absence of chlamydospores.

The production of chlamydospores by macroconidia of *F. solani* f. *phaseoli* has been elucidated by Nash, Christou & Snyder (1961), who found that when macroconidia were seeded into field soil, they either germinated to produce a short germ tube which eventually produced one or more chlamydospores, or else they were converted directly into chlamydospores. Chlamydospores thus represent the normal survival propagule derived from macroconidia. Nash *et al.* employed the technique devised by Warcup (1955) for determining the kind of propagule that has produced a colony on a soil dilution plate; by this method, they ascertained that most colonies of *F. solani* f. *phaseoli* arising on dilution plates made from soil of a naturally infested bean field originated from a chlamydospore, associated with or embedded in a fragment of plant tissue or of humus deriving from such a plant residue. Soil dilution plates made by Nash & Snyder (1962) from the soil of various infested bean fields, employing a selective peptone-PCNB agar, gave propagule counts of $1-3 \times 10^3$/g soil, and these propagules were found to be uniformly distributed through the soils down to the ploughing depth of 6–8 in. Ploughing and other cultivations quickly destroy any distribution pattern arising from that of the original production sites of pathogen propagules in the soil.

On account of the remarkable longevity of *F. oxysporum* f. *cubense* in soil in the absence of bananas or other host plants, much attention has recently been paid to the survival of this pathogen in the form of chlamydospores. The importance of these as survival propagules was clearly demonstrated by Newcombe (1960) in her study of the survival of this fungus on glass fibre tapes buried in soil. Failure of the fungus to survive in flooded soil was correlated by Newcombe with its failure to produce chlamydospores under anaerobic conditions. The oxygen requirement for production of chlamydospores by *F. roseum* f. *cerealis* has recently been demonstrated by Griffin (1968) by means of a new technique of exceptional elegance and promise. Very thin glass cells were filled up with glass microbeads (approx. 300 μ diam.) and the space between was partly filled up with nutrient solution, so as to give a growth medium simulating soil in its particulate nature and thus in its air–water régime but with the advantage of semi-transparency for direct microscopical observation. By this technique, Griffin was able to demonstrate an interesting difference between the air requirements for production of macroconidia and chlamydospores, respectively. Macroconidia were produced within water zones, though close to their air boundary; chlamydospores, in contrast, were produced only in the air spaces of this particulate medium. Pigmentation of mycelium was similarly restricted to that occupying air spaces, and this recalls the general observation that epiphytic mycelia of root-infecting fungi are often coloured, whereas infection hyphae, at least in the deeper layers of host tissue, are rarely so.

Much information was collected by Rishbeth (1955) about the survival of *F. oxysporum* f. *cubense* in the banana soils of Jamaica. Although Rishbeth mentioned the possibility that longevity of this pathogen was due to chlamydospores, he did not attempt to prove it, nor does this uncertainty detract from the persisting value of his extensive data; we can merely note that subsequent work makes it reasonably certain that his surviving propagules were chlamydospores. As I have already mentioned in Ch. 3 (p. 72), it is now accepted that *F. oxysporum* f. *cubense* cannot be identified with certainty from other forms of *F. oxysporum* on the agar plate, as Reinking & Manns (1933) had originally declared possible; pathogenicity to *Musa* species must be established before identification can be assured, laborious as such host-testing may be. For his pathogenicity test, Rishbeth employed small Gros Michel banana suckers, taken from a disease-free area and set in glasshouse pots filled with the soil to be tested. Rishbeth calibrated his technique by planting suckers in a series of disease-free soils inoculated with a suitable dosage range of macroconidia, and then counting mean number of rhizome infections per plant; values ranged from 0 to 6 and the test was able to detect infective propagules down to the low population level of 2 propagules/g soil.

F. oxysporum f. *cubense* was not detected by this technique in soils that had never carried bananas, nor usually on sites abandoned for bananas more than 8 years earlier. But a population below the detectable level for this technique, i.e. of less than 2 propagules/g soil, must usually have been present on such old banana sites, because when Gros Michel was planted, wilt usually developed before long. Indeed, Rishbeth quotes one instance in which only 2 banana plants out of 100 planted survived on a site that had not carried bananas for 20 years. In such cases, he was unable to detect any propagules of f. *cubense* until a few months after infected plants had wilted, when 50 propagules/g soil might be found, increasing in a year or two to a maximum of about 300 around diseased plants still standing. After removal of diseased plants, pathogen numbers declined after 2–3 years to a value around 10 propagules/g soil. But Rishbeth noted that if a diseased crop of Gros Michel was followed by a planting of the resistant Lacatan, instead of by fallow or a non-host crop, then this decline in population of *f. cubense* was slowed down; in one such case, the pathogen was still present at a level of some 100 propagules/g soil 9 years after planting Lacatan. Non-progressive infections of this disease-resistant banana evidently maintain the population level of the pathogen; in complete absence of bananas of any kind, Rishbeth observed that the pathogen population fell to approx. 3 % of its peak level within 3 years, and to below 0·5 % within 8 years.

Rishbeth's survey in Jamaica was followed by that made by Trujillo & Snyder (1963) of the banana soils of Honduras. The host for identification of f. *cubense* that they employed was *Musa balbisiana*, the advantage of which was brought to light by Stover & Waite (1960). Unlike Gros Michel, which sets seeds very rarely in its fruits (much to the detriment of dentists, as the seeds are large and excessively hard), *M. balbisiana* sets seed freely and so seedlings can be used for the host test; this saves not only labour but also the difficulty of finding disease-free areas from which to obtain healthy suckers. By the time of Trujillo & Snyder's investigation, it had become much more likely that chlamydospores constituted the survival propagules of f. *cubense* and they reported finding very large numbers of these chlamydospores in infected banana tissues. They found a high population of pathogen propagules only around sites of recently dead banana plants; this very local distribution of the pathogen was attributed to the absence of ploughing or other soil disturbance in banana plantations.

Following upon the original observations by Chinn (1953) noted above, it has now become well established that chlamydospores of various fusaria will germinate in response to a nutrient stimulus from substrates of fresh plant tissue and from root exudates from non-host as well as from host roots. This non-specific response of the chlamydospores, which might at first sight appear to be disadvantageous to survival of the population (as indeed it seems to be for the conidia of *Cochliobolus sativus*), actually enables them to augment their population. The fusarial chlamydospores germinate in the presence of the diffusing nutrients to form a brief mycelium on which more chlamydospores are produced; this ephemeral saprophytic phase results in a real increase in the volume of the chlamydospore population. Thus conidial chlamydospores of *F. solani* f. *phaseoli* studied by Schroth & Hendrix (1962) responded both to chopped fresh plant material of various kinds and to root exudates of at least some non-host species; in the rhizospheres of tomato, lettuce and maize, the chlamydospore population doubled itself. This saprophytic opportunism by f. *phaseoli* has been further confirmed by Toussoun, Patrick & Snyder (1963) and also by Papavizas, Adams & Lewis (1968).

Much of the recent work on these chlamydospores has been concerned with analysis of the nutrient composition of seed, root and hypocotyl exudates, and with the effects of these on chlamydospore behaviour. Toussoun & Snyder (1961) found that chlamydospores of *F. solani* f. *phaseoli* did not germinate spontaneously when air-dried soil was rewetted, but did germinate when placed in contact with bean seedling hypocotyls. Proximity to germinating seeds and to root tips was found by Schroth & Snyder (1961) to stimulate germination but the surface of mature roots was non-stimulating; this difference could be correlated with differences in amino acid and sugar content of the exudates. The following were detected in seed and young root exudates, and found individually to stimulate chlamydospore germination: aspartic and glutamic acids, asparagine, glucose and sucrose. Hypocotyl exudates contained sugars, but only trace amounts of amino acids. Schroth, Toussoun & Snyder (1963) identified twenty-two amino acids together with glucose, fructose, sucrose and maltose in exudates from bean seeds and roots. When added at the rate of 0·2 ml of a 2% solution to 0·5 g chlamydospore-infested soil, all four sugars and twenty of the amino acids individually stimulated germination to some extent, maximum responses being to glucose (46% germination) and to asparagine (42%). A combination of three sugars and seven amino acids at a total concentration of only 0·005% absorbed into porous porcelain seed models stimulated 12% chlamydospores to germinate, as compared with none in the water controls. When a combination of four sugars and four amino acids, in total amount equivalent to that in exudate from a single germinating bean seed, was dried down on an area of cover slip equal in area to the total surface of the seed, then 27% chlamydospores germinated; there was a 10% germination response to the sugar fraction alone, and one of 4% to the amino acid fraction alone. These responses were compared with 41% germination of chlamydospores in soil adjacent to germinating bean seeds, suggesting that sugars and amino acids together account for the major part of the germination stimulus from seed and root exudates. The relative germination response to sugars and amino acids, respectively, is likely to be affected by the available nitrogen content of the soil. In an alkaline loamy sand with low available nitrogen, Cook & Schroth (1965) obtained responses to single amino acids of three times the average response to single sugars; sugar and inorganic nitrogen in C/N ratio equivalent to that in asparagine gave a germination response equal to that elicited by asparagine. Penicillin and streptomycin much increased chlamydospore germination *in the absence of* exogenous nutrients, presumably by reducing either bacterial assimilation of spore-exudate nutrients or bacterial fungi static excretions or both (p. 178).

Reasons for failure of infection by chlamydospores after a high percentage of germination have been investigated by Cook & Snyder (1965), who attempted to discover why germinating bean seeds are rarely infected by f. *phaseoli*, whereas seedling hypocotyls are regularly infected in the same soil. This difference appeared to be due not to any difference in resistance to infection between seeds and hypocotyls, but rather to the effect of the respective exudates on chlamydospore behaviour. Thus 60% chlamydospores germinated within 16 h in soil in contact with germinating seeds, but growth of germ tubes then ceased and lysis shortly followed. In contrast, only 20% chlamydospores germinated within 48 h in soil adjacent to hypocotyls, but the germ tubes continued to grow, forming mycelia appressed to the hypocotyl surface and infection duly developed. Cook & Snyder suspected that this difference was due to the difference in nutrient composition of seed and hypocotyl exudates, respectively; in seed exudates, the sugars are balanced by amino acids (Schroth *et al.* 1963), whereas in hypocotyl exudates there are sugars but only a trace of amino acids (Schroth & Snyder, 1961). To produce a model for these two contrasting exudate effects, Cook & Snyder added carbon and nitrogen nutrients in various combinations to chlamydospore populations in unplanted soil. Glucose or sucrose alone stimulated only 20% of chlamydospores to germinate, but there was no lysis of the germ tubes; after a limited growth, the hyphae produced new chlamydospores. Both germination percentage and lysis of germ tubes increased when the nutrient solution contained nitrogen in balance with the sugar. Nitrogen as asparagine promoted earlier lysis than did inorganic nitrogen, and the lysis was maximal when yeast extract was added to a solution of glucose and asparagine.

Cook & Snyder have therefore concluded that ample provision of exogenous carbon and nitrogen nutrients together, either artificially or by way of seed exudates, promotes maximum germination of chlamydospores but simultaneously creates conditions that are fatal to the germlings, probably through the development of a large population of zymogenous bacteria. In support of this explanation, they quote a review by Brian (1960), who has carefully discriminated between various types of mycelial lysis. First, he has pointed out that autolysis in pure culture sets in only when carbon nutrients have become exhausted. Lysis caused by other soil micro-organisms is distinguished as heterolysis. But effective competition by a large microbial population can deprive a fungus germling of nutrients even when these are continuously renewed by seed or root exudation, and so induce a carbon-starvation type of autolysis similar to that eventually occurring in pure culture through carbon-substrate exhaustion. Other mechanisms of heterolysis are more easily distinguished from pure-culture auto-lysis. Thus Brian has noted that a large microbial population can cause oxygen-lack in spore germlings, and also that some anti-biotics act as respiration inhibitors and thus cause growth-stasis and death for the same ultimate reason. He further cites examples of bacterial and actinomycete enzymes that can disrupt fungal cell walls or protoplast membranes, so that heterolysis of fungal hyphae can act through a variety of mechanisms. Lloyd & Lockwood (1966) have also made observations on lysis of fungal sporelings and in their discussion have laid most emphasis on autolysis follow-ing unavailability or exhaustion of substrate nutrients. But the speed with which germ-tube lysis occurs in the situation examined by Cook & Snyder suggests that carbon starvation is not the only or even the most important cause.

These conclusions by Cook & Snyder were later employed by Cook & Flentje (1967) to interpret the behaviour of chlamydo-spores of *F. solani* f. *pisi* in a loamy sand soil (pH 7·5) in South Australia. Maximum germination (46%) of chlamydospores adjacent to germinating pea seeds occurred at the highest moisture content tested (81% saturation) but germling survival was better at lower moisture contents; in subsequent trials, percentage of germ tubes lysed increased with soil moisture content rising above 81% saturation. These results were interpreted by Cook & Flentje not as a direct effect of soil moisture content upon chlamydospore and germling behaviour, but rather as an indirect effect from the influence of soil moisture level on exudation by germinating pea seeds; exudation rate was shown (by dry weight loss of germinating pea seeds) to increase progressively with rising soil moisture content. A similar effect of rising soil moisture content upon rate of exudation by germinating pea seeds had earlier been reported by Kerr (1964) from the same laboratory (Ch. 2, p. 37). To test this interpretation, Cook & Flentje then added to their chlamydospore-containing soil a nutrient solution of 2·2% sucrose made up with ammonium sulphate to C/N 5; when any possible effect of soil moisture content on rate of nutrient accretion to the soil through seed exudation was thus blanketed, chlamydospore behaviour was not affected by soil moisture content over the range 58–90% saturation. After 17 h, some 53% chlamydospores had germinated and after 48 h some 35% were left with germ tubes still unlysed, irrespective of the particular soil moisture level. But addition of sucrose alone (2·2%) to the soil was not followed by any germ tube lysis, even in a soil at 100% saturation, so that all results were as predicted from the conclusion of Cook & Snyder about effects of exudate nutrients on chlamydospore behaviour.

Other observations on behaviour of chlamydospore populations in soils can possibly be interpreted by reference to these effects of carbon and nitrogen nutrients, and taking into account the modification of the C/N ratio of exudates by the nitrogen content of the soil solution. In general terms, we can expect that in a nutrient medium of low C/N ratio, carbon nutrients will be exhausted before available nitrogen, and so chlamydospore germlings will perish and autolyse through carbon starvation. But in a medium of high C/N ratio, exhaustion of nitrogen will terminate growth whilst carbon nutrients are still available, and so (in the absence of infectible host tissue) hyphae will continue to absorb surplus carbon nutrients and incorporate them into chlamydospore cell walls. Such a sequence of events could have determined the fate of macroconidia of *F. solani* f. *phaseoli* in two soils of the Columbia Basin region of the U.S.A. studied by Burke (1965). Both soils were of the same fine sandy loam type; one represented virgin soils recently brought under irrigation and very liable to severe outbreaks of bean root-rot; the other represented old irrigated soils of the same region, which had become 'resistant' to development of the disease. In his study of the behaviour of macroconidia in a film of water agar on glass slides buried in the soil, Burke found that the old, 'resistant' soil promoted extensive germ tube growth with a comparatively late development of small terminal chlamydospores on the mycelium. In the young, 'susceptible' soil, on the other hand, germ tube growth more quickly became

199

arrested and numerous large chlamydospores were produced both within the macroconidia and also terminally on short germ tubes. Burke's observations can be explained on the assumption that the old soil had a higher available nitrogen content than the young one; it is well known that in this type of virgin soil, both organic content and total nitrogen increase with continued cultivation of crops under irrigation. Thus in comparing nine virgin soils with three old irrigated soils in the same region of the U.S.A. as studied by Burke, Menzies (1959) found a mean increase in organic matter, attributable to irrigation and crop cultivation, of 53 %.

This account of the chlamydospores of *Fusarium* species, which have been described and figured by Toussoun & Nelson (1968), would be incomplete without some mention of the structure of the chlamydospore wall, which may determine differences in longevity between populations of different species and formae speciales. Differences in ultrastructure of the wall have been described by Nash & Alexander (1965) between chlamydospores of *F. solani* f. *cucurbitae* and f. *phaseoli*, respectively, and they suggest that these differences may account for the much longer life of chlamydospore populations of f. *phaseoli* in soil. This possibility has been further discussed by Nash Smith (syn. Nash, Shirley M.) in a review (1969).

Ascospores of Rhizina undulata

These ascospores are of exceptional interest, because they are the only ones known to me in which the dormancy-breaking effect of heat-shock, so widely employed in laboratory practice, has been demonstrated as a factor of paramount importance for initiation of outbreaks of a root disease. *R. undulata* is the cause of a root disease of conifers known as 'group dying', and outbreaks have been known for some years to be associated with bonfire sites, both in Europe and in the U.S.A. Conditions governing initiation of these outbreaks have been elucidated by Jalaluddin (1967*a*, *b*) for conifers in Great Britain. They arise only at the margin of bonfire sites, when branches from brashing, thinning or clear felling have been burned in close proximity to stumps of recently felled trees, or to standing trees of a very susceptible species. Although older trees can become infected, the greatest damage from the disease is suffered by young trees planted on a clear-felled site. Jalaluddin has demonstrated the sequence of events leading up to such an outbreak, as follows. During progress of a bonfire, the underlying soil and that on the perimeter is heated up; peak soil temperatures during progress of the bonfire are zoned and decrease centrifugally. A zone thus occurs below and around every bonfire in which the peak temperature lies within the range 35–45° C, determined by Jalaluddin as suitable for dormancy-breaking in the ascospores by heat-shock; the actual position of this zone, of course, will depend upon size and duration of the bonfire. Independently of its infection of tree seedlings, *R. undulata* manifests its presence by production of ascocarps on the soil surface. In the forest, conditions necessary for infection of tree seedlings and production of ascocarps are: (1) soil heating by a bonfire; (2) presence of susceptible roots, infection of which provides *R. undulata* with a food base from which to spread as mycelial strands through the soil, and thence to infect the roots of seedling trees; (3) an acid soil. Roots of recently felled trees are the most susceptible of all, for reasons explained in Ch. 4 (p. 86). If trunks of standing pine trees are close enough to the fire to be badly scorched, then root resistance may be reduced sufficiently to permit infection. Jalaluddin further observed that *R. undulata* could infect the roots of standing Norway spruce even when trees had suffered no fire damage, and Sitka spruce appears to be equally susceptible. Douglas fir roots, in contrast, appeared more resistant than those of pines. Roots of deciduous trees seemed to be completely resistant to infection by *R. undulata* even after trees had been felled; ascocarps did not appear around a bonfire site except in the presence of susceptible coniferous roots. The group-dying disease thus presents various features of exceptional interest to which Jalaluddin did full justice, particularly by providing quantitative estimates in the form of ascocarp counts. Table 14 taken from his second paper shows how he employed ascocarp counts to demonstrate: (*a*) that pine stump roots must be within reach of the mycelium of *R. undulata* arising from heat-shock-germinated ascospores so that it can infect them and thus establish itself in a first food base; (*b*) that roots of freshly produced stumps provide the best food base for *R. undulata*; with elapse of time from tree-felling, stump roots become first infected and later saprophytically colonized by various soil fungi, and are then no longer available for infection by *R. undulata*. Figures given in Table 14 are means of five replicates.

Table 14. *The effect on colonization of bonfire sites by* Rhizina undulata *of the position and age of pine stumps*

(From Jalaluddin, 1967*b*)

Mean no. ascocarps per fire site developing within 1 yr of bonfire

A. Position of fresh pine stump in relation to fire site

| Within fire site | Distance from margin of fire site (m) | | |
	1·5	3	6
20	13	5	0

B. Age of pine stumps near fire site

| Freshly felled | Period since felling (yr) | | |
	1	2	3
16	10	6	0

Jalaluddin (1967*a*) has also presented a variety of data on germination of the ascospores of *R. undulata* with and without a heat-shock treatment. Germination of unheated ascospores on 3% malt agar did not occur below 15° C; over the range 18–30° C, rate at which germination occurred increased with rise in temperature. Stimulation by heat-shock occurred over the range 35–45° C; the best treatment was 72 h at 37° C, followed by incubation at 22° C. In a typical experiment, 50% ascospores germinated between the 6th and the 8th day after heat-treatment and no more thereafter up to 30 days; only 10% untreated ascospores germinated within 30 days and germination was spread over the period from 8th to 29th day. These figures illustrate very well the point that I made early in this chapter, viz, that ascospores of this type will germinate without heat-shock but they do so slowly and irregularly over an extended period; heat-shock terminates endogenous dormancy quite abruptly for a substantial proportion of the spore population, so that they germinate quickly and together when environmental conditions are favourable. The peculiar interest of Jalaluddin's observations is that here nature is seen as appearing to imitate scientific art rather than vice versa.

In view of the well-marked endogenous dormancy of these ascospores, it is not surprising that Jalaluddin found them to survive well when buried on glass slides in three different forest soils. After 1 year, the number of spores found viable when slides were coated with malt agar, heat-treated and then incubated at 22° C, was about one-quarter of those that germinated when similarly treated at the outset. After 2 years, the proportion so germinating was reduced to 7% of the original in one soil, but none germinated in the other two.

Resting spores of the Plasmodiophorales

Resting spores of *Plasmodiophora brassicae*, causing the clubroot disease of crucifers, can survive for a number of years in the soil. A longevity of 5 years in New Zealand was recorded by Gibbs (1939), and Fedorintchik (1935) in the U.S.S.R. found 27% of cabbage seedlings to become infected when potted in soil taken from a field that had not carried a cruciferous crop for 7 years. A 4-year rotation is often inadequate to control this disease; on Agdell Field at the Rothamsted Experimental Station, which had been under a rotation of swedes with three non-cruciferous crops since 1848, severe outbreaks of clubroot eventually developed from about 1920 onwards. The disease is characterized by the development of large galls on the root system; the gall tissue eventually becomes transformed into a mass of resting spores, which are later liberated into the soil through decomposition of the gall tissue (Fig. 21).

When such a high population of resting spores is released into the soil following upon a severely diseased crop, then longevity of the pathogen for some years is assured, even if only a minute fraction of the original population preserves its dormancy. Germination of these resting spores is very pH dependent, and exogenous dormancy is probably more widely and effectively maintained by a sufficient degree of soil alkalinity than by any effect of soil fungistasis of microbial origin; this effect of soil

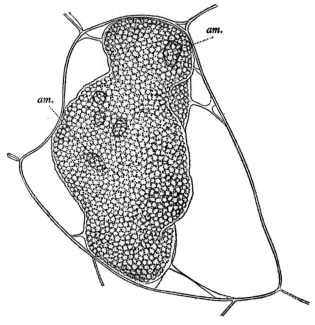

21. Formation of resting spores by *Plasmodiophora brassicae* in infected cell of host. (After M. S. Woronin, 1878.)

alkalinity is thus of the type classified by Dobbs & Gash (1965) as 'residual fungistasis'. The inhibiting effect of soil alkalinity operates not only in soil but also in pure aqueous solutions (Macfarlane, 1958). Under favourable pH conditions and in the presence of a cruciferous root exudate, the resting spore germinates to produce a single, biflagellate, primary zoospore, which has been described and electronmicrographed by Kole & Gielink (1962). Infection of the root hairs by these primary zoospores was described by Samuel & Garrett (1945), who have reviewed the earlier observations by Woronin (1878) and Cook & Schwartz (1930); the primary zoospore develops into a plasmodium within the root hair and within 7–8 days at 25° C this plasmodium becomes converted into a cluster of ripe zoosporangia, from which secondary zoospores emerge (Fig. 22).

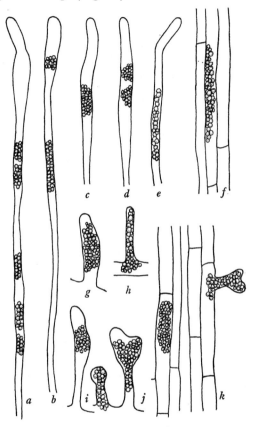

22. Formation of zoosporangia by *Plasmodiophora brassicae* in infected root hairs of cabbage seedlings. (After G. Samuel & S. D. Garrett, 1945.)

Employing an infected-root-hair-count method based on these observations, we found the optimum soil pH value for germination of resting spores and infection of root hairs was slightly above 6; an optimum pH range of 5–6 was found by Macfarlane (1958), employing the same root-hair-count method but in aqueous solutions instead of in soil or sand. Nevertheless, although the optimum pH value for root-hair infection lies within the pH range 5–6, alkalinity exercises a stronger inhibiting effect on spore germination than does a comparable degree of acidity, and this is one of the reasons for the general use of lime or ground chalk for control of clubroot in farm practice (Colhoun, 1953).

Of particular relevance to the longevity of resting-spore populations in soil are observations by various authors on wastage due to apparently spontaneous germination. To follow this downward drift with time in the resting-spore population, Bremer (1923, 1924) assessed by a plasmolysis technique the viability of spores in thin sections of galled cabbage roots buried under various soil conditions. A high proportion of spores were found to germinate spontaneously even in soil freshly treated with lime, although germination was more rapid and extensive in the acid, untreated soil. Similar results were reported by Macfarlane (1952), using the infected-root-hair-count method to estimate resting-spore populations; he found a rapid decline in populations during the first few weeks after addition to unplanted soil. This decline was more rapid in a wet, slightly acid soil, providing optimum conditions for spontaneous germination, than in a drier, alkaline soil. Similarly Bochow (1961) determined rates of spontaneous germination in three different soils of pH value 6·6, 7·0 and 7·2, respectively; in all three soils, more than 50% spores had germinated after 8–10 days, and more than 90% after 12 weeks.

In addition to this loss through spontaneous germination, resting-spore populations of *P. brassicae* in soil also suffer wastage through the germination-decoying effect of root exudates from some non-host species of crop plants and weeds. Webb (1949) was the first to report zoosporangia of *P. brassicae* in the root hairs of several grass species and other non-cruciferous plants; such plants can be termed 'non-hosts' in the sense that infection proceeds no further than the zoosporangial stage in the root hairs, and so no resting spores are produced. Macfarlane (1952) found root-hair infection not only in species from families closely related to the Cruciferae, such as *Papaver rhoeas* and *Reseda odorata*, but also in roots of some grass species, e.g. *Lolium perenne, Agrostis alba stolonifera*, and *Dactylis glomerata*. In one of his experiments, a crop of perennial ryegrass appeared to reduce the resting-spore population, by comparison with that surviving in fallow soil. Kole & Philipsen (1956) have similarly described zoosporangia in the root hairs of *Trifolium pratense*. It further appears that diffusates from some kinds of organic material also stimulate resting-spore germination, as demonstrated by Bochow & Seidel (1964) both for farmyard manure and for rye-straw meal.

A similar germination-decoy effect has been demonstrated by White (1954), who tested the response of various species in the Solanaceae to inoculation with resting spores of *Spongospora subterranea*, causing powdery scab of potato tubers. Amongst the species tested, *Datura stramonium* was the only one that stimulated germination of the spore-balls of *S. subterranea* without subsequent development of disease galls on the root system. In a field trial, plots planted with potatoes as a test crop showed a reduction of powdery scab from 37% infected tubers after a fallow to 7% following a thick planting of *D. stramonium* as a decoy crop.

SURVIVAL OF SCLEROTIA

Sclerotia are multicellular resting propagules, composed of vegetative hyphal cells, and they sometimes incorporate some host plant tissue. Though often irregular in shape, the constituent cells tend to be more or less isodiametric, thus differing from the elongated cells of septate hyphae. Sometimes, as in the sclerotia of *Rhizoctonia solani*, there is little differentiation between inner and outer cells; more often, however, the inner cells are thin-walled and colourless, and the sclerotium is bounded by a rind of cells, often of smaller size, with thicker, dark-coloured walls. Chemical analyses of various types of sclerotia have shown that the content of reserve nutrients, and especially of carbon nutrients, is usually much higher than that of unorganized mycelium.

A fundamental distinction can be made between sclerotia produced by air-borne fungal pathogens and those produced by root-infecting fungi. Sclerotia of the former group usually produce either fruit bodies, such as the stalked apothecia of *Sclerotinia* species, or stromata in which spores are produced, such as the stalked, perithecia-producing stromata of *Claviceps* species (ergot fungi of grasses and cereals). Sclerotia of this type are often variable in shape, and usually rather variable in size within a single species; they can attain a size of up to 2 cm in their largest dimension. In contrast with these are the sclerotia of most root-infecting fungi, which are both smaller and much more uniform in size and also more regular in shape, most frequently being subspherical. The sclerotia of *Phymatotrichum omnivorum*, causing Texas root-rot of cotton and many other dicotyledonous hosts, are mostly 1–2 mm in diameter, though occasionally reaching 1 cm (Rogers, 1937); they have been described as 'usually smaller than radish seed' by Taubenhaus & Ezekiel (1931). Both in size and shape, the sclerotia of *Sclerotium rolfsii* again resemble radish seed, being mostly 1–2 mm in diameter (Leach & Davey, 1938). The sclerotia of *Sclerotium cepivorum*, causing the white rot disease of onions and related species, are subspherical but smaller than those of *S. rolfsii*, being mostly 0·2–0·5 mm diameter (Coley-Smith, 1959). The microsclerotia of *Verticillium dahliae* are aptly named, because their range in length is given as 50–200 μ (Isaac, 1949).

As compared with sclerotia of air-borne fungal pathogens, sclerotia of root-infecting fungi are thus characterized by their smaller and more consistent size, and often by a greater regularity in shape (subspherical) as well. This smaller and more uniform size of sclerotia in root-infecting fungi can probably be related to the fact that they function individually as *single infective propagules*, producing on germination either hyphae or hyphal strands; sclerotia of air-borne pathogens, on the other hand, germinate to form either fruit bodies or stromata, each of which produces a large number of infective propagules (spores). As I pointed out in my discussion of inoculum potential in Ch. 1, all organisms, including fungi and higher plants, can be seen to exercise economy in the deployment of their reproductive resources; the most economical size for a reproductive propagule is the *minimum effective size*, in which definition some allowance must be made for a 'safety factor' to provide for at least a certain range of variation in the habitat, individual host plant or substrate of higher plant, parasite or saprophyte, respectively. The application of this generalization to the sclerotia of root-intecting fungi is quite clear; the most economical size is the minimum size that will generate an inoculum potential adequate for infection of the roots of the average or typical host plant. All the root-infecting fungi mentioned above have as host plants either herbaceous annuals or the smaller woody perennials; we can therefore conclude that the limited range in root-resistance to infection imposed by size-limitation in the host plants has imposed a similar size-limitation and uniformity on the sclerotia of the fungal pathogens. No such limitation is thus imposed on the size of sclerotia produced by air-borne pathogens, because they do not act as infective propagules directly; instead they produce a widely variable number of infective propagules, according to their size and reproductive capacity.

Finally, it is worth noting that fungi infecting tree roots do not usually produce sclerotia of the type defined above; it is the infected root that serves as a food base for a new infection. For several such fungi, Campbell (1933, 1934) has proposed the term 'pseudosclerotium' to designate a volume of infected wood enclosed by a membrane of sclerotized fungal cells; in Ch. 6 (p. 167) I have noted the differences between a pseudosclerotium and a sclerotium.

Phymatotrichum omnivorum

The morphogenesis of these sclerotia was first described by King & Loomis (1929), who showed that they arose at intervals along the mycelial strands (Fig. 16, p. 96) by a process of renewed cell growth and division, which occurred in the large cells of the wide central hypha as well as in the smaller cells of the narrower investing branch hyphae. This resulted in a central mass of closely packed, colourless, thin-walled cells of large and small size intermixed. The central mass, or medulla, is bounded by a rind composed of brown, thicker-walled cells of very irregular outline in surface view. From some of these surface cells, acicular hyphae (setae), similar to those surrounding newly formed mycelial strands, are produced (Fig. 23).

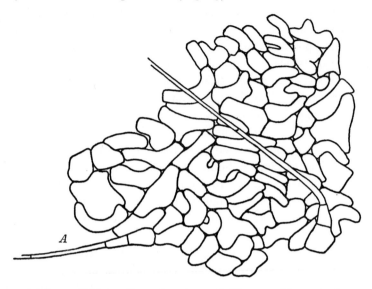

23. Surface cells bounding sclerotium of *Phymatotrichum omnivorum*. Acicular hypha at *A*. (After C. J. King & H. F. Loomis, 1929.)

During germination of the sclerotia, as described by Presley (1939), the new hyphae do not arise as extensions of the cell walls of the medullary cells, but are organized *inside* these cells, from which they later emerge. Ergle (1948) found that sclerotia taken from 30-day cultures of *P. omnivorum* would germinate on a sterile medium containing only inorganic salts, so that endogenous dormancy could have lasted only for a brief period; carbon reserves were in the form of glycogen.

In their study of sclerotium longevity, King & Eaton (1934) found that spontaneous germination in a well aerated soil contributed substantially to decline in the population of viable sclerotia. Evidence from field trials suggests that disturbance of the soil by cultivation implements also encourages spontaneous germination; cultivation temporarily improves soil aeration, though it may have another and more important effect on sclerotium populations, as we shall see shortly. But it is first necessary to appreciate the general sequence of events that determines the rise and fall of sclerotium populations in the field, as elucidated by Rogers (1937, 1942). Rogers's data show that formation of sclerotia occurs mainly after the conclusion of the parasitic phase of *P. omnivorum* on living host roots. Thus the highest population of viable sclerotia was found not at the end of the year under a susceptible crop (cotton) but at the end of the first year under a non-susceptible crop following it. Almost as many viable sclerotia were found after 10–12 years under non-susceptible grain crops as under continuously planted cotton. Rogers has pointed out that the number of sclerotia found at any time under a continuous sequence of susceptible crops is determined by the balance between rate of production on infected roots and rate of loss, through host-stimulated germination, of sclerotia produced in earlier seasons.

With Rogers's interpretation in mind, it is profitable to examine the data of Mitchell, Hooton & Clark (1941). Following upon a cotton crop with 75–95% plants killed by root-rot in 1939, a comparison was made between rotary cultivation on 30 October to a depth of 8–10 in, and normal ploughing later in the autumn as a preparation for the spring-sown crop of cotton in 1940. In August 1940, 36% plants were dead from root-rot on the rotavated plots, as compared with 90% on the ploughed plots. In the soil of the rotavated plots, no sclerotia were found down to 12 in, though they were present below this level. On the ploughed plots, as Mitchell *et al.* state, 'sclerotia were observed in abundance in the 6- to 12-in level'. Numbers of sclerotia thus present in August 1940 would have represented the number produced on the roots of the 1939 crop *minus* the number stimulated to germinate by the roots of the 1940 crop. Since the latter figure is unlikely to have been affected by the cultivation treatments in autumn 1939, it is evident that number of sclerotia *produced* after the two cultivation treatments must have determined the difference in numbers surviving at the sampling in August 1940. As Mitchell *et al.* have pointed out, early rotavation would have chopped up the cotton root systems and interfered with further extension of infection by *P. omnivorum* along the roots and with its production of sclerotia on the mycelial strands; ploughing later in the autumn would have interfered much less with the production of sclerotia. It is further possible that the soil-aerating effect of rotavation would have encouraged spontaneous germination of sclerotia already formed, though it seems likely that this would have been subsidiary in importance to the mechanical dismemberment of root systems by the rotavator blades. Rogers (1942) found a similar beneficial effect in reduction of root-rot to follow subsoil cultivation.

In the same field trial, Mitchell *et al.* also included as treatments the rotavation into soil of farmyard manure and sorghum fodder, respectively; these had the same effect on sclerotium populations in August 1940 as had rotavation alone, but reduced percentage of deaths from root-rot to a yet lower level. In accompanying laboratory experiments, Mitchell *et al.* showed that amendment of soil either with farmyard manure or with fresh plant material strongly stimulated germination of sclerotia. After germination, much the same sequence of events occurs as with some kinds of spore in the same situation; the sclerotia exhaust their nutrient reserves by outward growth of mycelial strands, but the strands are unable to secure a sufficient share of the germination-stimulating substrate to form new sclerotia and thus replace the population wastage so induced. Death of the young fungal colonies thus initiated can be ascribed to two causes, the relative importance of which is difficult to assess: (1) microbial competition for substrate nutrients, and especially for carbon nutrients, resulting in starvation and autolysis; (2) microbial lysis (heterolysis). As a result of other experiments similar to those in collaboration with Mitchell & Hooton, Clark (1942) stressed the heterolysis component of this effect, which he considered to be one of 'biological control'. Whether he was right in so doing is still debatable, but there is no doubt that the initiation of this sequence of events is due to a *direct* effect of substrate nutrients on the sclerotia, stimulating them to germinate; in current practice, biological control is defined as being mediated by one or more organisms (excepting man himself) outside the host–parasite relationship (Garrett, 1965).

Helicobasidium purpureum

This fungus causes the violet root-rot disease of sugar beet, carrot, legumes and various other field crops, and the sclerotia are more variable in size and shape than are those produced by other root-infecting fungi. They may develop on the mycelial strands, as described by Duggar (1915) and Faris (1921), but on crops with storage tap-roots, such as sugar beet, mangold and carrot, they are usually formed around small lateral roots; when sectioned, the vascular elements of the rootlet can be distinguished as the longitudinal axis of the sclerotium. Sclerotia formed thus around small lateral roots are elongated and often irregular and contorted in shape.

In the course of examining sugar beet and carrots with violet root-rot over more than 20 years, I have never been able to find sclerotia produced directly on the fleshy tap-root, though I could always find them on lateral rootlets and sometimes on the unswollen tail end of the tap root. This recalls an observation recorded by McNamara & Hooton (1933), who regularly found sclerotia of *Phymatotrichum omnivorum* on small lateral rootlets but not on the tap-root or larger lateral roots of infected cotton plants. It is most unlikely that the nutrients required for production of sclerotia are wholly supplied by the infected tissues of these small lateral roots, because the total nutrient content of the mature sclerotium must greatly exceed that of the original root before infection. It is probable that the rootlets provide a 'perch' on which sclerotia are produced by translocation through the mycelium or mycelial strands from the main body of infected tissue in the fleshy tap root. It is possible that the nutrient balance in the mycelium on the surface of the tap root is adverse to the initiation and maturation of sclerotia, but that this nutrient balance becomes altered to a favourable one as the mycelium travels away from the tap-root surface along the lateral roots.

This explanation if unsupported by other evidence could be dismissed as too speculative to commit to print, but it is possibly compatible with the results of a study I made on carbon and nitrogen requirements for production of sclerotia by *H. purpureum*

on nutrient agar (Garrett, 1949). Sclerotia are produced by
H. purpureum on suitable nutrient agars, but they vary widely in
size and also in position with respect to the colony centre; moreover, individual sclerotia often fuse during development to give
compound sclerotial masses. To overcome this difficulty, I
employed whole small colonies of *H. purpureum* as sclerotium-
survival units; ten colonies were grown around the periphery
of a 9 cm agar plate for an incubation period of 2 months at
25° C. After burial in soil for various periods, these small colonies
were exhumed and tested for presence of viable sclerotia by
inoculating small groups of young carrot plants. The basal nutrient
medium contained 1·5 % malt extract, providing vitamins and
approx. 1 % of sugars, and this was supplemented in different
treatment-series with 1, 2 and 4 % glucose, respectively; the
different sugar levels were combined in a factorial design with six
nitrogen levels over the range 0–0·16 %. Colonies were buried in
soil for a sufficient period (2 months) for elimination of
unorganized mycelium, so that survival could be equated with
production of mature, viable sclerotia; this was confirmed by
direct observation when colonies were exhumed. In this way, I
was able to show that production of sclerotia increased with rising
sugar content of the agar medium, up to the maximum concentration tested (5 %). At the lower sugar concentrations, nitrogen
above a certain critical level depressed production of viable
sclerotia; for 2 % sugar, this critical level was 0·01 % (C/N 80);
for 3 % sugar, it was 0·04 % (C/N 30); for 5 % sugar, it was not
reached at 0·16 % (C/N 12·5), which was the highest nitrogen
level tested. It thus appears that the C/N ratio of the medium does
not itself determine the initiation and maturation of viable
sclerotia, though one of its consequences may well do so. Thus at
any given sugar level, the nitrogen level will determine the amount
of sugar remaining in the medium after mycelial growth has used
up all the available nitrogen and has therefore come to a halt.
The amount of residual sugar will then determine whether sclerotia
are produced and if so, then the volume of sclerotial tissue formed.
It is now permissible, though admittedly speculative, to extra-
polate these results to interpret production of sclerotia on an

infected root system. Thus we can suppose that when *H. purpureum* is growing on a living root, vegetative growth of mycelium will continue until the available nitrogen, located chiefly in the cytoplasm and vacuoles of the host cells, is exhausted. After that, the mycelium will continue to hydrolyse both reserve starch and the carbon constituents of the plant cell walls until it has accumulated sugar to a level at which sclerotia are initiated and carried through to final maturation.

Spontaneous germination of these sclerotia was studied by Valder (1958), who employed sclerotial colonies on individual grains of rye as his survival units, and slices of living carrot-root tissue for his sclerotium-viability tests. He found that survival of the fungus on rye grains was reduced to a level of 10–40% in different experiments after some 4 months in a somewhat acid soil (pH approx. 5·3), and direct observation showed this to be due to spontaneous germination of the sclerotia. In a slightly alkaline soil (pH approx. 7·3), no spontaneous germination was observed, and 70–80% of the rye grains carried still viable sclerotia after 4 months in the soil. When the acid soil was brought to pH 7·3 by incubation with powdered calcium carbonate, then the sclerotia survived as well as in the other, originally alkaline soil, suggesting that soil pH itself directly controlled the occurrence of spontaneous germination. In neither soil, however, did addition of 1% of ground, dried plant material (peameal and grassmeal) stimulate sclerotium germination, because neither of these supplements significantly affected sclerotium survival in either soil. Both of Valder's results were somewhat unexpected, though stimulation of spontaneous germination by a suitable degree of soil acidity can be compared with a similar effect on the resting spores of *Plasmodiophora brassicae*, noted above.

Verticillium dahliae

V. dahliae produces microsclerotia in culture and on its host plants and can thereby be distinguished from *V. albo-atrum*, which develops dark-coloured mycelium but no microsclerotia in culture (Isaac, 1949). Most isolates of *V. dahliae* can further be distinguished from those of *V. albo-atrum* by their greater tolerance of growth temperatures above 23° C, and generally also by a higher optimum temperature for pathogenesis. European mycologists recognize these differences as worthy of specific rank, but in the U.S.A. both species are usually included within the single specific epithet of '*albo-atrum*'. The microsclerotia of *V. dahliae* are amongst the smallest known in root-infecting fungi; their range in length has been given as 50–200 μ by Isaac (1949), from whose paper Fig. 24 has been reproduced.

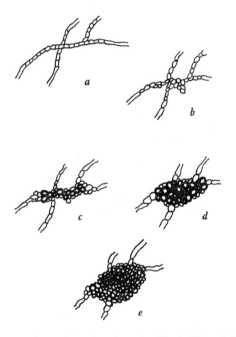

24. Development of a microsclerotium in *Verticillium dahliae*. Figures from *a* to *e* show growth stages over a period of 8 days. (After I. Isaac, 1949.)

The chemical composition of dried microsclerotia of age 4 weeks was determined by Gordee & Porter (1961) as 80% carbohydrate and about 6% each of lipid and protein; after 40 weeks more, carbohydrate content had fallen to just below 60%, whilst lipid had increased to 22% and protein to 8%. Microscopical examination showed the microsclerotia to be made up of melanized, thick-walled cells and hyaline, thin-walled cells in an apparently random mixture. When microsclerotia were sonically disintegrated and the tissue debris suspended in buffer solutions, germ tubes developed only from the thin-walled cells, more than 90% of which germinated over a wide pH range (4·2–9·2). Isaac & MacGarvie (1962) found that microsclerotia taken from 3–4 week cultures on Dox's agar would germinate on water agar, provided that they were first soaked in distilled water. Schreiber & Green (1963) observed that germination of the microsclerotia in soil was stimulated by root exudates, those from a host (tomato) being more effective than those from a non-host (wheat). They also confirmed the observations of Gordee & Porter on the morphology of the micro-sclerotia, and on restriction of germination to the thin-walled cells.

Because of the minute size of these microsclerotia, their fate in soil cannot be followed by the methods employed for larger sclerotia, i.e. those of 1–2 mm or more in diameter; thus soil populations of the sclerotia of *Phymatotrichum omnivorum* can be estimated by washing infested soil through a bank of sieves, as employed by Rogers (1936). The fate of individual sclerotia of this size or larger can also be followed by burying them against glass slides or within nylon envelopes. But none of these methods is practicable for the microsclerotia of *V. dahliae*, and so Isaac (1946) made an ingenious use of 'hyaline variants' of this fungus that did not produce microsclerotia. Agar colonies of the wild type survived for at least 5 months in the soil, whereas those of hyaline variants were short-lived. Schreiber & Green (1962) found that homogenized microsclerotial cultures (4 weeks old) survived in soil (0·5 mg inoculum/g soil) for at least 82 weeks; in contrast, homogenized young cultures (4–6 days) lacking microsclerotia failed to survive 21 weeks in soil, even at the highest inoculum level (7 mg/g soil).

Sclerotium rolfsii

Under some conditions, the sclerotia of *S. rolfsii* will germinate spontaneously, in the absence of any nutrient stimulus from fresh plant tissue or from root exudates. This was incidentally demonstrated by Leach & Davey (1938) by their viability test for sclerotia separated from samples of field soil by washing over a bank of sieves. The washed sclerotia were plated out in Petri dishes on the surface of moist, unsterilized, black peat soil and their germination was so regular that this technique provided a much more reliable viability test than did plating out surface-sterilized sclerotia on potato-dextrose agar.

In my earlier account of the behaviour of sclerotia (Garrett, 1956a, p. 184), I noted an interesting but quite isolated observation by Dunlap (1943) on production of sclerotia by *Phymatotrichum omnivorum*. He found that this could be completely inhibited in autoclaved Houston black clay soil, which normally supported a good production of sclerotia, by autoclaving 325-mesh dusting sulphur with the soil, at the rate of 1 part sulphur to 1000 parts air-dry soil. If the mixture was sterilized dry, no restraint on sclerotium production eventuated. But mycelial growth through the autoclaved soil–sulphur mixture was as good as that through soil alone. Recent work on production of sclerotia by *Sclerotium rolfsii* has now suggested an explanation for Dunlap's observation; Chet, Henis & Mitchell (1966) have reported that production of sclerotia by this fungus on nutrient agar was inhibited by five sulphur-containing amino acids and also by glutathione over the range 10^{-5}–10^{-4} M, without affecting the mycelial growth. Although not restricting mycelial growth, all these substances induced production of basidia of the perfect state of *S. rolfsii*, i.e. *Corticium rolfsii*, which had never been seen in culture before, and only rarely on its host plants. This morphogenetic effect of L-cysteine was found to be competitively antagonized by iodoacetic acid; at a molar ratio of 30 L-cysteine:1 iodoacetic acid, the converse effects of these two morphogens cancelled one another out, and growth of *S. rolfsii* was normal. The effect of iodoacetic acid by itself, over the range 3×10^{-5}–10^{-4} M, was to induce precocious production of sclerotia, which grew to larger than normal size. The morphogenetic action of the inducer was found to be limited to hyphal tips at the colony margin.

Following up this discovery, Chet & Henis (1968) found that

218

radioautograms of *S. rolfsii* grown on ^{14}C iodoacetic acid showed a selective accumulation of radioactivity first on future sites of sclerotium production and subsequently in the developing sclerotia. In contrast, radioactivity of ^{14}C L-cysteine was quite evenly distributed through the mycelium and in the few sclerotia that were formed. So the high molar ratio (30:1) of L-cysteine required to neutralize the morphogenetic activity of iodoacetic acid could be explained by its even distribution throughout the mycelium, whereas that of iodoacetic acid was more economical, being restricted to sites of sclerotium development. Chet & Henis further reported that chelating agents and potassium iodate, in addition to iodoacetic acid, induced production of sclerotia; they have suggested that all these inducers operate by modifying a repressor, which they think to be a sulphydryl-containing, copper-linked protein entity. They regard the action of L-cysteine as one of increasing either rate of production or of activity of this repressor, or of slowing down the rate of its turnover.

It seems likely that similar morphogenetic effects will be demonstrated with *Phymatotrichum omnivorum*, and perhaps with some other sclerotium-producing fungi as well. Any method for repressing production of sclerotia on infected crop plants that can be developed as an economic proposition for agricultural use certainly deserves further exploration.

Sclerotium cepivorum

In its host range, *S. cepivorum* is restricted to species of *Allium*. The behaviour of its sclerotia also conforms with the view that it is a highly specialized parasite, because they germinate freely in natural soil only when stimulated by root exudates of *Allium* species, or by volatile effluents from these plants. Coley-Smith (1959) conducted a survival-test with the sclerotia (0·2–0·5 mm diam.) buried in soil in nylon-mesh bags outside in the field; the percentage of viable sclerotia, determined by ability to germinate on malt-extract agar, after 4 years burial in soil was 70, 75 and 80 at depths in soil of 3, 6 and 9 in, respectively. Percentage infectivity, estimated by inoculation of onion seedling roots, was

100 after 3 years and 75 after 4 years. In some experiments, addition of fresh organic material to the soil reduced the population of sclerotia, but in others it failed to do so. In his next paper, Coley-Smith (1960) reported that sclerotia taken directly from 6-week-old cultures were endogenously dormant, inasmuch as they failed to infect onion seedlings. This endogenous dormancy began to be broken after 1 month's burial in soil, had nearly disappeared after 2 months and had completely vanished after 3 months; it could be broken by abrading the sclerotial rinds but, not surprisingly, this treatment impaired infectivity. Germination was stimulated both by root exudates and by root extracts of *Allium* species, but not by those of non-host plants. Percentage germination of sclerotia adjacent to onion roots was highest amongst those in contact with the apical region, which is known to be most active in exudation (Ch. 2, p. 36). The mode of germination by these sclerotia has been figured by Coley-Smith (Fig. 25).

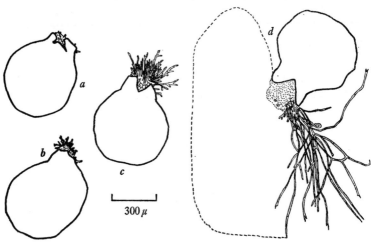

25. Stages in germination of the sclerotium of *Sclerotium cepivorum*, showing rupture of the rind (*a*), followed by extrusion of a plug of mycelium (*b*, *c*, *d*). (After J. R. Coley-Smith, 1960.)

This conclusion concerning the specificity of germination-response by the sclerotia to root exudates and extracts of *Allium* species was confirmed in a more extensive investigation by Coley-

Smith & Holt (1966), who tested sixty-three plant species from a wide range of families. Germination-response to root exudates of non-hosts did not exceed the maximum percentage of spontaneous germination (14%) shown by the particular isolate of *S. cepivorum* employed in these tests. In a comparative test with eleven fungal isolates, percentage spontaneous germination showed a general correlation with percentage germination-response, indicating that isolates varied in the degree to which exogenous dormancy was enforced by soil fungistasis. Response to 1% sugar ± nitrogen as casein hydrolysate in natural soil did not significantly exceed percentage spontaneous germination. Comparable tests with a number of nutrient solutions and plant extracts, but under sterile conditions on agar, silica gel, filter paper and sterilized soil, were made by Coley-Smith, King, Dickinson & Holt (1967); under these conditions, germination response was both high and quite unspecific, as has been the general experience with fungal propagules of many species on a wide range of nutrient media.

In discussing possible mechanisms for this highly specific response of the sclerotia in natural soil to root exudates of *Allium* species, Coley-Smith & Holt selected two as most worthy of further exploration. First, it was possible that the germination-activator was resistant to decomposition by soil micro-organisms, implying that it was not a nutrient substance. Secondly, such an effect could possibly be produced by an antibiotic to which *S. cepivorum* was much more tolerant than were other soil micro-organisms, so that the germination-stimulating nutrients in the root exudate were protected from competitive assimilation by other soil micro-organisms and hence were preserved for the sclerotia of *S. cepivorum*. This second hypothesis was thus much the same as that advanced by Cook & Schroth (1965) to explain the germination-response of chlamydospores of *Fusarium solani* f. *phaseoli* in soil to antibacterial antibiotics. Evidence for or against this second hypothesis was sought by Coley-Smith, Dickinson, King & Holt (1968), who appear to have eliminated its application to the sclerotia of *S. cepivorum*. They found that germination of the sclerotia in natural soil was induced by *Allium* extracts at a concentration too dilute to exercise any antibiotic effect on a number of common soil bacteria that they tested; they further failed to find any evidence for the view that *undamaged* roots of *Allium* species exert any significant antibiotic effect on the rhizosphere microflora. Previous claims for such an effect were ascribed to insufficient precautions against damage to roots under investigation; injured roots do excrete antibiotics at a significant concentration. Coley-Smith *et al.* therefore concluded that they were left with the first of the two hypotheses proposed by Coley-Smith & Holt, viz. that the germination-activator is a substance resistant to microbial decomposition.

The most recent advance in this investigation has been reported by King & Coley-Smith (1968), who found that sclerotia were induced to germinate in natural soil by volatile effluents from growing onion and leek seedlings, and also by air bubbled through diluted onion and garlic juices and by distillates and distillands of garlic juice. They also noted that diallyl sulphide, which possesses little antibiotic activity, was stimulatory, and suggest that this and other germination-activators produce a germination-stimulus sufficiently strong to override the effect of soil fungistasis. This explanation is compatible with all the evidence on behaviour of these sclerotia, which appear to be the most highly specialized and efficient of any yet investigated.

8 Principles of root-disease control

In this concluding chapter, I shall discuss the application to practice of the fundamental knowledge with which I have been dealing in preceding chapters. Plant pathology is a science based on agricultural practice, and all root-disease investigators must wish to see the results of their work eventuating in increased crop yields. For this, a thorough familiarity with the ecology and plant husbandry of the particular crop is absolutely essential; for root-disease control, appropriate modifications or innovations in crop husbandry practice have furnished the foundations for success in all types of crop cultivation. The man who finally has to make a disease-control method work in practice is the farmer, the glass-house grower, the plantation manager or the forester. In addition to controlling diseases, he has also to control insect pests, nematodes and weeds, and to follow a crop-husbandry régime that will give him maximum yield and sometimes maximum quality of produce as well; finally, he has to fit all these often conflicting demands into a seasonal programme in which labour requirements are spread out as evenly as possible. The final arbiter of the success of any new control method is the cost, and this censor has obliterated a good many ingenious ideas that seemed promising on paper. The ideal new control method is one that gives an increase in crop yield even in the absence of disease, because few farmers will knowingly plant a crop under the threat of a serious disease outbreak; heavy crop losses from root disease are much more often the outcome of ignorance rather than the result of taking a calculated risk. By the use of such a method, disease control becomes a bonus for good crop husbandry. Methods of this type can indeed be found, but only by the plant pathologist who takes trouble to understand *all* the grower's problems, and not merely those with which he is himself directly concerned.

In the discussion of control methods shortly to follow, I shall restrict myself to those that are particular to diseases of the root

223

system and stem base. The most obvious example of a control method that is general to the whole of plant pathology is the production of disease-resistant cultivars by the plant breeder. Production of a crop variety that remains resistant to disease, though many do not because of evolution or selection of new biotypes of the pathogen, represents the ideal control method, since the first cost is also the last. Here, however, the root-disease pathologist can often render a special service to the plant breeder, in devising a satisfactory technique for testing disease-resistance of new crop selections. As I have recounted in a brief history of root-disease investigation (Garrett, 1956a, Ch. I), a formidable obstacle encountered by root-disease pioneers was the task of reproducing the disease by inoculation; frequent failures can now be ascribed to an inoculum potential inadequate to establish infection. An equal but opposite danger has been the use of too high an inoculum potential, which has overridden any restrictive effect of soil conditions or varietal resistance that would have shown up under a less severe test and might have proved useful for disease control. No plant breeder needs to be told that the best trial ground for new crop selections is a naturally infested field. But such natural trial grounds in convenient proximity are often hard to find, and so the root-disease pathologist can here prove a useful ancillary to the plant breeder by providing a substitute for the natural arena.

Crop husbandry practices vary widely with the type of crop, but a convenient division can be made into field crops, intensive crops, and plantation crops. Field crops are usually annual, though a few are biennial, e.g. sugar beet for seed production. Intensive crops, whether grown under glass or out in the open, are usually demi-annual, i.e. two or more crops are raised in succession during a single year. Plantation crops are woody perennials, either bushy or arborescent, and may sometimes stand for 50 years or more, so that problems of root-disease control differ widely from those in the cultivation of field or intensive crops.

ROOT DISEASES OF FIELD CROPS

Soil conditions, and to some extent subaerial conditions also, affect root-infecting fungi in all the various phases of their life-cycle, and they can also modify disease-resistance or determine disease-escape of the host plant, thus:

(1) effects on parasitic phase of the fungus during growing season of the crop;

(2) effects on disease-resistance or disease-escape of the crop plant;

(3) effects on saprophytic or dormant survival of the fungus in absence of a host crop, or of weed host plants;

(4) effects on dispersal of the fungus, through air-borne or splash-borne dispersal of spores above ground, and through carriage of spores by mass-movement of water above or within the soil.

When considering the various phases in the life-cycle of a fungal pathogen, it seems logical to begin with infection of the host and to proceed in the sequence I have given above; the tendency to do this has indeed been reinforced by generations of textbook writers. But when we come to consider the application of environmental effects to disease control, it becomes more logical to take these various phases of the fungal life-cycle in the reverse order. Thus the first efforts of the plant pathologist must be directed towards securing pathogen-free soil for sowing of the crop, though in actual practice 'pathogen-free' means soil in which the pathogen has been reduced to an acceptable level rather than to zero. Traditionally, this has been secured by the universal practice of *crop rotation*, assisted by ancillary measures of *crop sanitation*. In the past, the traditional practice of crop rotation was often flouted in newly developed agricultural areas for so long as crop monoculture of wheat, cotton, etc., paid an immediate return, though at a considerable ultimate cost in the loss of soil fertility, particularly through soil erosion by wind and water. Tradition in this respect is still being flouted today but more safely, because more farmers have realized that it is bad business to allow their most important capital asset, i.e. a fertile soil, to depreciate. When the plant pathologist, in consultation with the farmer and his general agricultural advisers, has done what he can to reduce the soil population of the pathogen to an acceptable level, his task is still not completed. He must next devise *palliative measures* that will slow down infection and spread by the pathogen over the crop root system during the growing season, and perhaps also augment disease resistance or promote disease escape.

Crop rotation

The origins of crop rotation in the fairly distant past are the concern of the agricultural historian, and plant pathologists will find much to interest them in a history of agriculture by Gras (1940). The efficacy of crop rotation in controlling most of our important soil-borne diseases depends on the fact that they are caused by specialized pathogens. Thus monoculture of a particular crop promotes the development and continuity of a soil population of one or more specialized pathogens, such as *Ophiobolus graminis* on wheat and *Phymatotrichum omnivorum* on cotton. Conversely, in a good crop rotation in which the majority of crops are taxonomically remote from one another so that they do not share the same specialized pathogens, the sequence of hosts is broken by one or more consecutive years of non-hosts, so that the specialized pathogens are starved out of the soil. Thus the host range of *O. graminis* is restricted to species in the Gramineae and that of *P. omnivorum* does not include any monocotyledons, so that for control of either of these pathogens the farmer is left with adequate latitude in the choice of other crops for the rotation. Successful control of specialized root-infecting fungi by crop rotation gives a stronger assurance than any of the laboratory experiments I described in Ch. 5 that saprophytic multiplication on dead crop residues is of negligible importance in the life-cycle of these specialized pathogens. Nevertheless, crop rotation has its limitations. It is much less effective for the control of pathogens that produce long-lived populations of resting propagules (Ch. 7). The value of the best crop rotation can also be destroyed by inefficient farming that permits weed host plants, or volunteer hosts self-sown from earlier crops, to flourish unchecked.

Some effects of crop rotation or of its converse, crop monoculture, are still imperfectly understood. In general theory, any
pathogen specialized to a particular crop should continue to cause
fairly widespread and severe disease in that crop for so long as
monoculture continues. But there are exceptions of various kinds
to this general prognosis. Thus McNamara, Hooton & Porter
(1931) made yearly maps of cotton root-rot patches in Texas, and
found that such disease patches might disappear altogether from
their original sites after several more years under continuous
cotton and be replaced by a stand of apparently healthy plants.
Similarly, Fellows & Ficke (1934) reported that a take-all patch
appearing in a run of consecutive wheat crops in Kansas generally
increased in size during its second year but often decreased in its
third year and might not reappear at all after that. This sequence
of events can perhaps be explained by the fact that a fungal
pathogen survives less long on the root systems of small, early
killed plants in the centre of a disease patch than it does on the
larger root systems and stem bases of plants on the periphery of
the patch that are infected relatively late in the growing season.
Secondly, the early killing of plants in the centre of the patch may
result in a partial fallow if the surface soil is dry or in a green
manure crop of weeds if it is moist; in either case, nutrients are
conserved in the soil or returned to it, with the result that soil
fertility is likely to be higher within than outside the site of the
former disease patch when the next crop is sown. This enhanced
fertility on the site of the former disease patch is likely to decrease
the incidence both of take-all (Stumbo, Gainey & Clark, 1942) and
of cotton root rot (Jordan, Nelson & Adams, 1939; Adams, Wilson,
Hessler & Ergle, 1939; Blank, 1944).

A parellel phenomenon still in need of explanation is the gradual
decline of the take-all disease over whole fields, and indeed over
much larger areas, during monoculture of wheat. When wheat
monoculture, or its less extreme alternative of intensive wheat
cropping, has been begun on a virgin soil of low organic content
and correspondingly low fertility, this decline in incidence of take-
all can probably be explained by a gradual increase in soil organic
content and fertility under a satisfactory farming régime; an

15-2

increased level of soil nutrients promotes disease escape of the wheat plant (Ch. 1, p. 6) and a higher microbial population may exert an increased degree of biological control of *Ophiobolus graminis*. Thus Perkins (1917) has described the gradual decline of take-all after some 25–30 years from first clearing of the land from bush scrub on light, sandy soils of originally low fertility in various regions of South Australia. A similar sequence of events, though more compressed in time, has been described by Bosma (1946, 1962) for the newly drained polders on the area of the former Ysselmeer in the Netherlands. In the second to the fourth consecutive wheat crop on these polders, severe outbreaks of take-all have occurred but thereafter the disease has declined. Much of this decline can probably be ascribed to an increase under wheat cultivation of organic content, fertility and microbial population in a lake-bottom deposit originally very poor in all these respects. The increase in microbial population during the transition of the original lake-bottom deposit into a cultivated soil is certainly likely to result in increased biological control of *O. graminis*; this fungus spreads more rapidly over the wheat root system in soils that have been partially sterilized, and this is true for most fungal pathogens. Gerlagh (1968) has studied the microbiological aspects of take-all decline on these polder soils; he was able to produce a similar effect in glasshouse culture of wheat by reinoculating the soil every 3 months with fresh inoculum of *O. graminis*. Similar evidence for development of an increased level of microbial antagonism towards *O. graminis* has been presented by Lester & Shipton (1967) in England, where take-all decline sometimes occurs in a consecutive run of wheat crops on old arable soils. The microbial antagonism thus developing is postulated as being specifically induced by wheat monoculture; I must confess that I find it difficult even to visualize a possible mechanism for such an effect, though it would be rash to deny the possibility of its occurrence. Nevertheless, even on these old arable soils, other factors may contribute to take-all decline. Thus from their survey of fertilizer practice amongst English farmers, Rosser & Chadburn (1968) quote figures demonstrating a gradual rise in the level of nitrogen applications during a run of consecutive wheat crops from

64 units/acre for the first crop to 102 units/acre for the fourth and fifth. English farmers practising intensive wheat cropping are only too well aware of the threat of take-all and endeavour to balance increasing disease risk by a corresponding increase in the prophylactic dose of nitrogen, as the figures produced by Rosser & Chadburn clearly demonstrate. In England, nitrogen is the nutrient most likely to be in short supply; in the early days of wheat farming in South Australia, a shortage of phosphate constituted the major soil deficiency, and spectacular responses to superphosphate both in increased crop yield and in a decreased incidence of take-all were commonly obtained (Samuel, 1934).

The gradual decline of take-all with transition from a virgin soil of low fertility to an arable soil of higher fertility is paralleled by the behaviour of potato scab, caused by *Streptomyces scabies*, in a low-rainfall region of central Washington in the U.S.A. Menzies (1959) has described the gradual transition from the badly scabbed potatoes grown on land recently brought under irrigation to the almost scab-free crops grown on old irrigated fields. His figures show a decline in soil pH value and an increase in soil organic content with age of fields from first cultivation; both these changes would tend to reduce the incidence of scab, though an associated change in the soil microflora may be the most important one, as Menzies has suggested.

A final point that needs to be stressed about crop rotation is that the length of the break between susceptible crops required to reduce the pathogen population to a safe level varies within fairly wide limits even for a single species of pathogen. First, the longevity of the pathogen population increases with its size, which is determined by the volume of infected residues left behind by a diseased crop. Secondly, longevity of a population of any given size will vary with soil type and treatment, as described in Chs. 6 and 7. On a soil particularly favourable for underground spread of the pathogen during the growing season, a pathogen population of any given size will both produce a higher incidence of disease and leave more infective propagules behind after harvest than on a soil less favourable for infection and spread of the pathogen. For this reason, an application of lime to the soil is likely to reduce the length of rotation required for control of clubroot of crucifers, but to lengthen that needed against take-all of cereals and Texas root rot of cotton.

Crop sanitation

Under this heading come all those control measures designed to prevent dispersal of a pathogen through infected seed, plant propagating material, and crop residues, or by means of infective propagules carried by wind, rain-splash, moving water or animals. Infective propagules of root-infecting fungi can be carried by all the agencies, physical, human and animal, that distribute infective propagules of plant pathogens in general, and so there is no need to elaborate on this further here. I will, however, mention a mode of dispersal of novel interest, and that is by the cutting blades of the mowing machine used for harvesting forage crops; the blades serve very efficiently for the transmission and inoculation of infective propagules of vascular wilt pathogens. This was first suggested as a means of dispersal for *Verticillium albo-atrum* in crops of lucerne (*Medicago sativa*, known as 'alfalfa' in the U.S.A.) by Isaac (1957). This suggestion was later verified by Heale & Isaac (1963).

Also under the heading of crop sanitation can be comprised all those control measures that can be taken to hasten the disappearance of pathogen populations, whether ensconced in decaying crop residues or lying free within the soil as dormant propagules. There are two ways of dealing with this problem: first by removal or by destruction *in situ* of infected crop residues after harvest, and secondly by crop husbandry practices designed to hasten the natural disappearance of the pathogen from such residues lying in the soil. Bodily removal of plant remains, roots and all, from the soil has not found favour as a practical expedient, though it has been practiced as a way of controlling virus leaf curl of cotton in the Sudan (Massey, 1934). In the earlier days of wheat farming in Australia, the long straw left after harvesting by the wheat stripper was usually burned, and this helped to control both take-all and flag smut (caused by *Urocystis tritici*); later, however, this practice fell into disfavour on areas of light sandy soils liable to severe wind erosion in the long dry summer of South Australia. It now seems possible, however, that the same end can be achieved, but without the wasteful destruction of this source of valuable soil organic matter, by the recently introduced practice of *direct drilling* of wheat in stubble land without prior ploughing or any other cultivation; before drilling, weeds are destroyed by a herbicide spray. Brooks & Dawson (1968) have described a number of field trials in which plots ploughed in the conventional way before preparing a seed-bed were compared with direct-drilled plots. In the latter, wheat was drilled directly into stubble ground, or into a grass sward, and weeds were controlled by spraying with the herbicide paraquat (formulation 'Gramoxone' W). In all these trials, direct drilling gave substantial reductions in the incidence of take-all, as compared with that developing on the ploughed control plots. The occurrence of any direct fungicidal action of paraquat on *Ophiobolus graminis* was excluded by the fact that incidence of take-all on plots that were sprayed before ploughing was just as high as that on plots ploughed without herbicide-spraying. One important feature of direct drilling into stubble ground is that the infected tiller-bases of take-all plants are left above ground and hence *out of reach of the roots of the following wheat crop*. These infected tiller-bases constitute the most substantial and decomposition-resistant part of the wheat stubble left behind after harvest, and so their isolation from the roots of the next planting may well be responsible for much of the reduction in incidence of take-all that follows the practice of direct drilling. Brooks & Dawson have commented on this possibility, but also think that soil conditions under the direct-drilled plots are less favourable, from mid-April onwards, for underground spread of infections by *O. graminis* than they are under the conventionally ploughed plots.

There remain for discussion those crop-husbandry measures that the farmer can employ for hastening disappearance of a pathogen from the soil after harvesting a diseased crop, some of which have already been discussed, or at least implied, in Chs. 6 and 7 on saprophytic and dormant survival, respectively. The chief expedients open to the farmer are different modes of cultivation, special use of chemical fertilizers and the growing of green manure crops. I have already discussed in Ch. 6 the paramount effect of soil nitrogen level on longevity of saprophytic survival in several cereal foot-rot fungi. Soil nitrogen level can be quickly and easily raised by application of a nitrogenous fertilizer; it can be lowered by incorporation of crop residues of high C/N ratio. One way of doing this is by ploughing in a catch crop grown for the purpose and allowed to grow to a sufficient age before doing so; the C/N ratio of plant tissue increases progressively with age of the crop. We ourselves have advocated the use of an undersown legume or legume–grass mixture for control of take-all under a run of consecutive, spring-sown barley or wheat crops (Garrett & Buddin, 1947; Garrett & Mann, 1948). After harvest of the cereal, the undersown crop makes a vigorous growth if soil moisture permits, and quickly absorbs available soil nitrogen, thus depriving *O. graminis* of this nutrient essential for its saprophytic survival. Although legumes are thought of as crops that enrich the soil with nitrogen, it is a fact that when actively growing under good light conditions, they absorb more nitrogen from the soil than they excrete into it. There are two advantages of using a legume, or a legume–grass mixture, as an undersown crop for controlling take-all. First, a legume makes a quicker get-away after cereal harvest, when available soil nitrogen is temporarily low. Secondly, the nitrogen fixed by the legume is released by decomposition of its tissues after ploughing into the soil in mid-winter; it becomes available for plant uptake by the time that the next spring-sown cereal is ready for it, and this nitrogen then assists the cereal to withstand the effects of infection by *O. graminis* (Ch. 1, p. 6). An alternative method for hastening disappearance of *O. graminis* from infected stubble, tested by Scott (1969), has given promising results. After harvest, the infected stubble was shallowly rotavated into the soil, so as to produce a straw–soil compost with a proportion of straw much higher than when stubble is ploughed into the soil. We can assume that the aeration in the straw–soil compost produced by rotavation would be better than that around straw ploughed into the soil; secondly, the higher proportion of straw to soil in the rotavated compost would create a local scarcity of available nitrogen. Both of these effects would tend to accelerate disappearance of *O. graminis* from the infected stubble (Ch. 6); Scott's data suggest, however, that the effect of nitrogen scarcity was subordinate to that of improved aeration.

Some kinds of organic material, and especially green manures, when incorporated with the soil will stimulate a proportion of resting propagules to germinate in response to nutrients and other germination-activators diffusing out from them into the soil (Ch. 7). Such an effect may partly, or even largely, explain the success obtained by King (1937) in his use of heavy organic manuring for control of Texas root-rot of cotton in Arizona. Such a germination-stimulating effect of organic material on dormant propagules of root-infecting fungi is a *direct effect*, i.e. it is not mediated by other soil micro-organisms growing upon the organic material and therefore cannot be described as 'biological control'. In the early days of enthusiasm for biological control, almost any alleviation of a soil-borne disease by organic manuring was not unnaturally ascribed to some mechanism of this sort; Clark (1942) and his associates (Stumbo, Gainey & Clark, 1942) performed a timely service by demonstrating that release of nitrogen through the decomposition of nitrogen-rich organic materials might reduce incidence of disease through a direct manurial effect on the crop plant, an effect that could not be described as biological control at all. It is further essential to bear in mind that organic manuring does not assist in control of all root-infecting fungi; unspecialized pathogens like *Sclerotium rolfsii* are able very readily to colonize fresh plant tissues, and these are best kept out of the way in a soil heavily infested with this fungus (Ch. 2, p. 49).

A mechanism of biological control through microbial heterolysis of fungal cell walls has been postulated in consequence of a finding by Mitchell & Alexander (1962) that root-rot of bean could be reduced by manuring soil naturally infested by *Fusarium solani* f. *phaseoli* with chitin at 500 lb/acre. Actinomycete numbers were greatly increased following addition of chitin, but not those of bacteria, whilst those of fungi were somewhat depressed. Pursuing the hypothesis that addition of chitin and related substances to the soil promotes heterolysis of chitin-walled fungi, Mitchell &

Alexander (1963) reported their detection of laminarin, or a poly-saccharide closely related to it, in the hyphal walls of *F. solani* f. *phaseoli*, but not in those of *Pythium de baryanum*. In parallel with this finding, Mitchell (1963) reported that soil amendment with either chitin or laminarin at 500 lb/acre reduced incidence of bean root-rot and also of radish wilt due to *Fusarium oxysporum* Fr. f. *conglutinans* (Wollenw.) Snyder & Hansen. But neither amendment reduced diseases caused by two pathogens that are known not to contain chitin in their cell walls: *Pythium de baryanum* and *Agrobacterium tumefaciens* (E. F. Sm. & Towns.) Conn. Diffusates from soil incubated for 2 weeks with either chitin or laminarin at 500 lb/acre inhibited growth of the two *Fusarium* pathogens and lysed preformed mycelium; the diffusates had no effect on *P. de baryanum*. Encouraging results in control of bean root-rot followed amendments of infested soil with either ground lobster shell (12–20% chitin) at 250–750 lb/acre or with tissue of the seaweed *Laminaria* at 1 000 lb/acre.

Palliative measures to hinder spread of infection

Although, as I have just remarked, it is difficult to sort out the various effects of fresh organic manures on soil-borne diseases, a good example of an effect on the process of infection has possibly been provided by Snyder, Schroth & Christou (1959), working with *Fusarium solani* f. *phaseoli*; they found that bean foot- and root-rot could be controlled by organic amendments of high C/N ratio, such as wheat and barley straw (C/N 80) and corn stover (C/N 60), whereas amendments of low C/N ratio, such as soybean and alfalfa residues (C/N 16), made the disease worse. These effects can perhaps be interpreted through the finding by Toussoun, Nash & Snyder (1960) that exogenous nitrogen is essential for infection of the bean hypocotyl by propagules of *F. solani* f. *phaseoli*. These results have been confirmed by Maurer & Baker (1965), who have ascribed successful disease control by amendments of high C/N ratio to a locking-up of available soil nitrogen in the cells of micro-organisms decomposing the organic material. This is thought to deprive chlamydospores of the pathogen of even the minimal amount of soluble nitrogen required for germination and host infection.

The farmer has more control through crop husbandry practices over some other soil factors than might at first be supposed. Soil temperature cannot be changed, but time of sowing can be. In crops grown under natural rainfall, cultivation can do something towards moisture conservation; in some areas, crop irrigation or sprinkling can provide direct control of soil moisture content. Most root-infecting fungi are favoured by good soil aeration; the skilful use of cultivation implements at the right time can do much to consolidate the seed-bed, and great importance was attached to this requirement for control of take-all by pioneer wheat farmers in South Australia (Griffiths, 1933). Good control of diseases favoured by acid soils, such as vascular wilts caused by formae speciales of *Fusarium oxysporum* and clubroot of crucifers, can sometimes be obtained by liming the soil. For other crops in the rotation, however, an overdose of lime is likely to be harmful, and lime should therefore be applied in moderate doses at sufficiently frequent intervals. In addition to aggravating the incidence of such diseases as potato scab, Texas root-rot of cotton and take-all of cereals, an overdose of lime can render soil phosphate less available for plant uptake, and similarly take out of circulation minor elements like boron and manganese; uptake of most mineral nutrients is favoured by a soil pH around 6. Nutrient deficiencies, especially of minor elements, are often the primary cause of diseases of the root system associated with infection by otherwise weakly parasitic fungi (Ch. 2, p. 50).

ROOT DISEASES OF INTENSIVE CROPS

The term 'intensive' as used here implies a system of crop cultivation either under glass or outside in the field, in which both capital investment and costs of labour and materials are unusually high and bring a correspondingly high return from the crop produce. The opposite of this is extensive farming, as exemplified by the original pioneer farmers in North America and Australia; capital and labour were both in short supply and low investment brought only low returns per acre, though it is not customary to assess the virtues of pioneers by the criterion of economics alone. Intensive crop cultivation has reached its highest development within the glasshouse industry. The most advanced glasshouse growers have already achieved satisfactory control, sometimes fully automated, of the following environmental factors: temperature, light intensity and duration, atmospheric humidity and gaseous composition (i.e. by enrichment with carbon dioxide), and aeration, moisture content and nutrient composition of the rooting medium. Many years ago, glasshouse growers abandoned natural soil as a rooting medium and replaced it by composts made from natural soil mixed in various proportions with peat, quarry sand and fertilizers. Quite apart from the fact that supplies of natural soil from building sites and elsewhere became inadequate to supply the growing demand, these artificial composts could be varied to suit the special requirements of individual crops. From their own experimental work, root-disease workers know that much more precise control of the soil environment can be obtained with rooting media more artificial than soils and composts; examples are quartz sand, vermiculite, aluminium oxide grits and aerated nutrient solutions. Just so, no doubt, in the glasshouse industry of the future; soil organic matter and the microbial population it supports will become an unnecessary and unwanted complication. Unnecessary, because the plant nutrients provided by organic matter in natural soil can be more efficiently supplied and controlled by addition of fertilizers to a microbiologically inert medium; unwanted, because fungal and other pathogens that gain access to partially sterilized soil or compost can multiply therein more rapidly than in natural soil, where there is a higher degree of biological control by the resident microbial population. The use of microbiologically inert rooting media at the outset partially solves this problem, but only partially, because accumulating root residues from crops will tend to create something like a soil unless they can regularly be eliminated. This difficulty is avoided by growing crops in aerated nutrient solution, the composition of which can be controlled by continuous circulation and monitoring. Commercial application of solution culture was pioneered under the name of 'hydroponics' by

Gericke (1940), and has been further promoted by Sholto Douglas (1959) and others.

At the present time, however, many glasshouse growers are still using composts as the rooting medium. Growers of intensive crops in the open field employ soils originally natural, though subsequently much modified by heavy additions of organic manures and chemical fertilizers. Economic restriction to high-value crops, several of which are usually grown on the same ground in a single year, limits the range of crop choice and thus makes unworkable the control of disease through crop rotation. The place of crop rotation as a basic control measure has been taken by partial sterilization of the soil at regular intervals. By 'partial' sterilization is meant a treatment by heat or chemicals that is sufficient to eliminate pathogens. A degree of soil sterilization more severe than this is not only unnecessary but also undesirable, because it is important to preserve, as far as possible, the biological control of pathogens provided by the microbial population of natural soil. In the glasshouse industry, soil steaming for the eradication of pathogens, insect pests and weed seeds has been effectively practised for many years, but has suffered from the disadvantage that all organisms incapable of surviving exposure to 100° C are also killed; any pathogen recolonizing a recently steamed soil can spread at first with much reduced restriction by microbial competition, and so can cause more disease than in a similar but untreated soil. The disadvantage of soil steaming is thus that it raises soil temperature to a higher level than is really necessary. An ingenious method for controlling the temperature of heat treatment was first suggested by Morris (1954) in England, through the use of steam–air mixtures; by controlling the composition of the steam–air mixture, any desired temperature can be maintained. Application of this method to glasshouse practice was pioneered by Baker & Olsen (1960) in California and by Baker (1962) again in Australia. In addition to conservation of much of the soil microbial population, the steam–air mixtures are cheaper than pure steam, and post-steaming toxicity of treated soil to seedlings and delicate plants is avoided or much reduced. Maintenance of soil temperature by a steam–air mixture at 60° C for 30 minutes is sufficient to kill most fungal pathogens; Dawson, Johnson, Adams & Last (1965) found that *Rhizoctonia solani* reintroduced into soil pre-heated to 60° C caused less damage to seedlings than in soil pre-heated to 100° C. Baker, Flentje, Olsen & Stretton (1967) reported that vigour of saprophytic growth by three isolates of *R. solani* in several soils increased with the temperature to which soil was preheated, over the range of 49–100° C tested.

When soil is heated to 100° C, the population of nitrifying bacteria is usually reduced much more severely than that of the ammonifiers; a consequence of this is a temporary accumulation of ammonia in soils after steaming. After a marathon investigation, Johnson (1919) concluded that the chief cause of post-steaming toxicity to seedlings was this accumulation of ammonia. Data more recently but less copiously produced by Dawson *et al.* (1965, 1967) suggest that this ammonia-toxicity is augmented by toxic levels of manganese and nitrite. Lawrence & Newell (1936) recommended addition of superphosphate to steamed soil as a means of reducing this toxicity; it is now possible, however, by the use of these steam–air mixtures to prevent its development in the first place.

For intensive crops raised in the field, soil fumigation with volatile chemicals replaces heat treatment as the best method for elimination of pathogens. Fumigants are self-dispersing after injection into the soil and thus have a great advantage over non-volatile fungicides, which have to be applied in a water drench, or else mixed dry with the soil. Future use of such non-volatile chemicals is likely to be restricted to regions where too low a temperature or too high a soil moisture content restrict dispersion of fumigants. Grainger (1959) has devised a rotavating machine for mixing fungicidal and nematocidal dusts evenly with the top 9 in of soil. Under favourable conditions, however, soil fumigation has now no rival as a practical method for elimination of soil-borne fungal, bacterial and actinomycete pathogens, nematodes, insect pests and weed seeds. These favourable conditions are found par excellence in areas of low rainfall, such as California, where relatively high soil temperatures and the close control of soil moisture content possible under irrigation practice can provide optimum conditions for fumigant dispersion (Goring, 1967, 1969). Speed of dispersion increases with rising soil temperature. A medium soil moisture content is optimum for fumigation; if the soil is too wet, dispersion is restricted by water-filled pore spaces, whereas in too dry a soil, pathogens are more resistant to the action of fumigants and of lethal agents in general. In California, S. Wilhelm estimated in 1965 that some 3500 acres of strawberry, tomato and vegetable land were fumigated annually; both the widespread use of fumigants and the high pitch of technical development in fumigation practice owe much to the pioneering work of Wilhelm and his associates (Wilhelm, Storkan, Sagen & Carpenter, 1959; Wilhelm, Storkan & Sagen, 1961; Johnson, Holland, Paulus & Wilhelm, 1962). Fumigants are injected with a tractor-drawn, automatic injector and the soil is covered for 24–48 h after fumigation with a polyethylene tarpaulin (20 × 1400 ft). For control of Verticillium wilt of strawberries on heavily infested land, fumigation with chloropicrin at the rate of 480 lb/acre has been recommended by Johnson *et al.* (1962). But for general soil sanitation, in the absence of a major soil-borne pathogen or pest, Californian growers employ a mixture of chloropicrin and methyl bromide (in the ratios 55:45 or 33:67 by weight) at 250–400 lb/acre. Wilhelm *et al.* (1961) have recorded strawberry yields in fumigated soil of nearly four times the average yield in untreated soil.

The very consistent and sometimes spectacular increases in crop growth that follow fumigation, even in the absence of any major pathogen or pest, are similar to those following partial sterilization by steam; speculation about the mechanism of this effect has been going on since the end of the last century, and particularly since the major investigation by Russell & Hutchinson (1909) at the Rothamsted Experimental Station. Three microbiological changes in the soil in consequence of partial sterilization by heat or by fumigants probably account for most of the observed increase in crop growth. First, as I have already noted, partial sterilization temporarily inhibits or slows down the process of nitrification without affecting ammonification, so that ammonia accumulates. Tam & Clark (1943) attributed the faster growth of pineapple plants after soil treatment with steam, formalin, or chloropicrin to the fact that ammonia is a better source of nitrogen than is nitrate for this particular species. The same seems to be true for seedlings of at least some coniferous species, and especially for those of Sitka spruce. Such a preference for ammonium nitrogen may have contributed to the spectacular growth responses of coniferous seedlings to soil fumigation which has been reported by Benzian (1965) from a very extensive and long continued series of experiments in forest nurseries carried out in Great Britain by Rothamsted Experimental Station in collaboration with the Forestry Commission.

Partial sterilization brings about a change not only in the form but also in the total quantity of soluble soil nitrogen, and this second change is probably the more important of the two for most plant species. As I have already pointed out, partial sterilization of soil kills off most of the microbial population, as well as most of the soil fauna; death and autolysis of microbial and other cells quickly releases nitrogen, phosphate and other nutrients locked up in the protoplasts, and more slowly those fixed in cell walls. This rapid increase in soluble nitrogen, phosphate and other nutrients can account for much of the crop growth response to partial sterilization. But it is essential to make one point quite clear; partial sterilization cannot increase the *total* soil content of nitrogen or of any other plant nutrient. It merely releases, in a single event, the nutrients locked up in living cells; in the normal sequence of nature these would be released in due course, but much more gradually. It is evident, therefore, that this moiety of the total benefit of partial sterilization is likely to be a wasting asset; continued fumigation of soil from which nutrients are regularly carried off as crop produce will lead to soil impoverishment unless the drain of plant nutrients is balanced by adequate replenishment with fertilizers.

Realization of this fact that nutritional benefits from partial sterilization are finite and exhaustible leads to the conclusion that they cannot account for the maintained increase in crop yields that often seems to accompany the continued practice of regular soil fumigation in the absence of any major pathogen or pest. This may well be due to a third effect of soil fumigation, i.e. the recurrent elimination, within fumigant range, of a whole spectrum of some less obvious pathogens and pests. Individually, no species may do much damage, but collectively they destroy much of the plant root system and particularly the part most important in nutrient absorption, i.e. the small and usually ephemeral feeding rootlets (Wilhelm & Nelson, 1969). This was the conclusion reached by J. P. Martin and his associates (Aldrich & Martin, 1952; Martin, Aldrich, Murphy & Bradford, 1953), as a result of their extensive investigation into fumigation of old citrus soils in California. This is also the explanation favoured by S. Wilhelm and his associates, in the papers cited above.

ROOT DISEASES OF PLANTATION CROPS

The life of a plantation crop extends over a period of years, and in the case of some tree crops can extend to 50 years and more. During this period, pathogenic root-infecting fungi can continue to spread over the root systems and from infected to healthy trees in root contact, though host resistance generally increases with age of the bush or tree; progress of infection is thereby slowed down and may eventually be halted altogether. With plantation crops like fruit trees in temperate regions and rubber, cacao, tea and coffee in the tropics, the value of the product is high enough to justify regular application of fertilizers and routine spraying with fungicides and insecticides against foliage diseases and pests, which practices tend to maintain crops in good general health and thus to promote resistance to root diseases. With forest plantation crops, however, the value of the timber on an annual yield assessment is low by comparison with the annual return from a horticultural plantation crop and so expenditure on maintenance has to be reduced accordingly. For the same reason, forest plantations are usually restricted to soils, sites and climates that are unsuitable for other crops, though often giving good yields of timber. Plantation growers and foresters, however, are alike faced by the same general problem; the long life of their crops makes it difficult to check the spread of any root-disease outbreak, once it has started. There are three obvious and general ways of reducing losses from root disease. First, particular attention must be paid to choice of soil, site and climate for a particular crop; the more closely that environmental conditions approximate to the ecological optimum for the crop species, the greater will be its resistance to diseases and pests. Secondly, every effort must be made to ensure a clean start for the young crop on land as free as possible from infected root systems left behind by the preceding crop or forest. Thirdly, crop husbandry practices must be so designed as to exclude subsequent entry of root-disease pathogens into an originally healthy plantation. Disease control measures and crop husbandry practices vary widely according to type of crop and geographical region. Despite this wide variation, I can illustrate the most important features of current control régimes by brief accounts of two important root diseases in which good progress has been made towards satisfactory control; these are the diseases caused by *Fomes annosus* in Great Britain and by *F. lignosus* in Malaya, respectively.

Fomes annosus *on conifers in Great Britain*

Control measures in practice or under development against the root- and butt-rot diseases of various conifers caused by *F. annosus* can all be comprised within the general term of crop sanitation, as defined above. First, there are those control measures designed to prevent infection of the still living stumps of recently felled trees by air-borne spores of *F. annosus*. Secondly, control measures are required for dealing with infected stump root systems left behind in the soil after clear felling of one rotation crop in preparation for planting of the next.

Rishbeth (1951 a) was the first to demonstrate conclusively by inoculation experiments that air-borne basidiospores consistently infected the freshly exposed wood surface of pine stumps resulting from tree-felling. Both at that time and subsequently, he accumulated a convincing mass of evidence that air-borne infection of stump surfaces provides the means whereby *F. annosus* eventually establishes itself in infected roots in the soil of plantations originally free from this pathogen, as when trees were planted on land that had been under agricultural crops for many years previously.

The living tissues of the tree stump can be infected only by parasitic fungi able to overcome residual host resistance; spore infection of freshly exposed stump surfaces is thus confined to *F. annosus* and to a few other species like *Peniophora gigantea* that are weak parasites but not sylvicultural pathogens (Ch. 6, p. 166). As Rishbeth originally demonstrated, the mycelium of *F. annosus* developing from basidiospore infections grows from the surface down into the massive body of the stump and thence out along the stump roots; for some time after felling, the stump roots remain alive and residual host resistance is still sufficient to exclude invasion by obligately saprophytic fungi, though not sufficient to check appreciably the rapid progress of *F. annosus* (Ch. 4, p. 87). In addition to basidiospores formed on the *Fomes*-type fruit bodies, *F. annosus* also produces asexual spores of *Oedocephalum* type as its imperfect state. The *Oedocephalum* conidiophores and conidia, which are produced only when atmospheric humidity is near 100%, are highly distinctive in appearance and have been extensively employed by Rishbeth as a diagnostic character for detection of the mycelium of *F. annosus* in woody tissues, after prior incubation in a moist chamber. In Great Britain, where the climate is suitable for basidiospore production over much of the year, it is doubtful whether the *Oedocephalum* conidia contribute much to effective dispersal of the pathogen. This view is strengthened by the data of Kuhlman & Hendrix (1964) in the U.S.A. on the relative infectivities of populations of basidiospores and conidia, respectively; a suspension of basidiospores at a concentration of 14/ml was equal in infectivity to a conidial suspension of 8600/ml. Nevertheless, Bega (1963) has expressed the opinion that the conidia are more important than the basidiospores in the forests of California, where production of basidiomycete fruit bodies is greatly restricted by the virtual absence of summer rainfall.

It thus became clear from Rishbeth's early work on this disease that felling of trees carried with it a high risk of air-borne stump infection by *F. annosus*. In sylvicultural practice, trees are commonly felled at various times in addition to the final clear felling for harvest of the timber. Thus strips of trees are felled through young plantations to provide 'racks' for access and 'rides' for fire prevention. At their original planting, seedling trees are set closer together than in the final spacing that is optimum for their growth to maturity. This allows for mortality from various causes, and a close early spacing encourages a straight vertical growth of the young trees and discourages development of large branches; both effects increase the ultimate value of the trunks for timber. The developing plantations are therefore usually thinned more than once; for species of pine grown in East Anglia, the first thinning is usually made when the plantation has reached an age of some 20 years. The first control measure advocated by Rishbeth and adopted by the Forestry Commission as a routine practice for East Anglia was the painting of stump surfaces with creosote immediately after felling (see Rishbeth (1967) for a historical account of this and subsequent developments). Creosote as a stump protectant provided a useful stop-gap measure for a number of years, but its effectiveness was limited by the fact that it provided no more than a passive, fungistatic seal against infection (Rishbeth, 1959a). Within this seal, the stump tissues remained alive for longer than in stumps left untreated and thus exposed to infection by the fungus air spora. When the protective seal of creosote was broken either by desiccation-cracking of the underlying wood or by mechanical damage by wheeled traffic, then the wood was just as liable as it was immediately after felling to infection by basidiospores of *F. annosus*. To overcome this weakness of the creosote treatment, Rishbeth (1959b) sought for chemical treatments that would give a more active protection than the purely passive type provided by creosote. Chemicals found most effective as active protectants were ammonium sulphamate, urea and certain boron compounds, such as disodium octaborate, mostly as 20% solutions. These solutions penetrated the stump to some depth from the surface and killed the tissues; by thus destroying residual host resistance, they enabled a wide range of saprophytic fungi to invade the stump by spore colonization of the surface and thus to compete with *F. annosus* for colonization of the substrate. The mechanism of stump protection against *F. annosus* provided by these chemicals thus seems to operate chiefly through their phytocidal effect, which opens the way for biological control of *F. annosus* by various competing fungal saprophytes. Biological control thus mediated by a chemical treatment can be compared to biological control of *Armillaria mellea* by species of *Trichoderma* as mediated by soil fumigants, evidence for which I have discussed in Ch. 6 (p. 169). The method employed against *F. annosus*, however, is both more effective and more consistent in its action. As Rishbeth has pointed out, early colonization of stump tissues by saprophytic or weakly parasitic fungi has an advantage additional to that of stump-surface protection; the downwards progress of harmless fungal invaders from the stump surface into and along the stump roots is likely to arrest upwards invasion of the stump system by *F. annosus* coming from any part of the root system that it had succeeded in infecting *before* the tree was felled. In further screening tests by one of Rishbeth's research associates (Punter, 1963), sodium nitrite was found to be superior as an active protectant to any chemical hitherto tested; after further trials by the Forestry Commission Research Branch, this has been adopted as the standard treatment of this type. Sodium nitrite is toxic to mammals and needs safe handling; in forest areas open to access by unattended stock, or in those serving as a catchment for water reservoirs, sodium nitrite is replaced by urea or disodium octaborate, or by the biological control method next to be described. In the south-eastern U.S.A., borax is widely used as a stump protectant for pines (Driver, 1963).

As early as 1952, Rishbeth reported experiments in which pine stumps were inoculated with oidia (asexual spores) of *Peniophora gigantea* in an attempt to control infection by basidiospores of *F. annosus*. The results of these pilot experiments were promising but the project must have seemed to outside observers, as it did to me at that time, as one likely to go the way of earlier attempts at biological control by inoculation, which were overcome by the technical difficulties of devising a sufficiently cheap and fool-proof method for general commercial use. Nevertheless, Rishbeth persisted and eventually produced a viable method, which is the first example of successful biological control through inoculation to have been adopted in plant pathological practice, so far as I know. In describing this method, Rishbeth (1963) has pointed out that he originally selected *P. gigantea* for this purpose because it is the commonest and most effective competitor of *F. annosus* in the natural colonization of pine stump surfaces; it is a vigorous decomposer of lignin as well as of cellulose, and the mycelium established through spore infection of the stump surface grows quickly into the body of the stump and thence into the stump roots. Moreover, *P. gigantea* can often do more than merely halt advance by *F. annosus* along the stump roots from root infections established before felling; sometimes it is able to continue its advance through sections of root already occupied by *F. annosus*, which it thus replaces (Ch. 6, p. 165).

Rishbeth determined that a dose of 1×10^4 oidia of *Peniophora gigantea* allowed a satisfactory margin of safety for effective biological control of the largest spore inoculum of *F. annosus* likely to be naturally deposited on a pine stump of 16 cm wood diameter. He solved the technical problems of storage and application of the spore inoculum by the production of dehydrated tablets containing approx. 1×10^7 viable oidia, and these were found to have a storage life of at least 2 months at $22°$ C. Further technical improvements in the production of inoculum are on the way, but the success of the prototype is attested by present employment of it by the Forestry Commission over some 70000 acres (Rishbeth, 1967). The method is suitable for stump inoculation of pines which are the trees chiefly grown over the low-rainfall area of East Anglia; the most important species are Scots pine (*Pinus sylvestris*) and Corsican pine (*Pinus nigra* var. *calabrica*). Present efforts are being concentrated on finding suitable fungal competitors for protection of stumps of other coniferous species, such as Sitka spruce (*Picea sitchensis*), larch (*Larix decidua*), Douglas fir (*Tsuga heterophylla*) and *Thuja plicata*; these are the principal conifers grown in the north and west of Britain. Whereas losses in pine species due to *F. annosus* are caused chiefly by tree killing, in the other coniferous species just cited the main loss is due to development of butt rot (Ch. 4, p. 89).

There can be no doubt that Rishbeth was right in giving the

first priority in his research programme to this search for methods of preventing effective dispersal by *F. annosus*, and especially for keeping it out of plantations established on healthy sites; the problem of dealing with this pathogen once it has become established within infected roots on any plantation site is much more difficult and correspondingly more costly. Nor is the importance of taking these preventive measures limited to areas in which *F. annosus* fruit bodies occur. By sampling the fungus spora of the air over Scalloway in the Shetland Islands, Rishbeth (1959*c*) was able to detect spores of *F. annosus* at a concentration that was about 3% of the maximum occurring over the East Anglican pine plantations near Thetford; fruit bodies nearest to Scalloway were probably about 180 miles away on the mainland of north-east Scotland.

The second priority in research towards control of *F. annosus* has been given to the quest for some method of dealing with infected stump roots left behind after clear felling of a plantation, in preparation for establishment of another plantation on the site. At one time, this appeared to be an intractable problem, at least where the population of infected roots left behind after clearing was likely to be high. One possibility would be to admit defeat and to follow with a coniferous species chosen not for its suitability to the site but for its possession of a useful degree of resistance to *F. annosus*; such a species is *Abies grandis*. The present outlook on this problem, however, is somewhat more hopeful, as results of experiments on mechanical extraction of stump root systems initiated by the Forestry Commission Research Branch are beginning to come in. The cost of extracting the old root systems, by winching and/or lifting by suitably designed machines, appears to be less than might have been anticipated, so that cost of extraction is likely to be more than balanced by the increase of timber yield from the first crop alone. Interim figures from the results of these experiments have been quoted by Rishbeth (1967), in a review of control measures against *F. annosus* in Great Britain.

Fomes lignosus *on rubber in Malaya*

In order to understand the evolution of the present scheme of control measures against white root disease, the most serious root disease of rubber in Malaya, it is necessary to know something about the outstanding pioneering work of R. P. N. Napper at the Rubber Research Institute of Malaya over the decade 1930–40; this work was cut short by his untimely death as a war casualty. Shortly after the description by R. Leach (1937, 1939) of his successful ring-barking method for control of *Armillaria mellea* in East Africa, Napper (1940) reported early results from field trials of stump poisoning with sodium arsenite. Stump poisoning was designed to achieve exactly the same objective as that attained by ring-barking well in advance of felling, i.e. to kill the root system as quickly as possible so that it is soon invaded by saprophytic fungi; this halts any further advance by *F. lignosus* along roots in which it is already established, or with which it is in contact, at the time of felling. Stump (or tree) poisoning, thus designed to prevent spread of *F. lignosus* over stump root systems after felling, has been successfully applied in Malaya in conjunction with clearing land either from jungle or from an old rubber plantation, in preparation for replanting with rubber. This effective method for reducing incidence of white root disease in young rubber plantations is still employed in Malaya, though sodium arsenite, a dangerous poison to mammals, has been replaced by the arboricide 2,4,5-trichloro-phenoxyacetic acid (2,4,5-T).

Napper (1932, 1938) also developed another control method against white root disease; this was designed to deal with any roots infected by *F. lignosus* that were present in the soil at time of planting. For the first few years in the life of a young plantation, periodical inspections were made of the collar region of every young tree by removing the soil from around the upper part of the tap root and the proximal part of the surface laterals; the rhizo-morphs of *F. lignosus* are epiphytic for several feet in advance of root invasion, so that their detection in the collar region gave an early warning of infection advancing from outlying parts of the root system, before the young tree had suffered irreparable damage.

When *F. lignosus* was thus discovered, the rhizomorphs were traced back to their infection source, part of an old stump root system; this was then extracted from the soil and destroyed, together with sections of the young rubber roots that had actually become infected. Purely epiphytic rhizomorphs deprived of their food base are harmless, and these were merely pulled away from the root surface and removed; hindsight from later work now enables us to say that the further practice of swabbing the root surface with a 2% solution of copper sulphate after 'derhizomorphization' probably did nothing to increase the effectiveness of the control procedure, but almost certainly did quite a lot for the professional morale of the pest-gang labourers.

This control method devised by Napper was widely practised by commercial rubber estates in Malaya for a number of years, but rising labour costs, amongst other factors, eventually made it imperative to design a cheaper procedure towards the same end. This procedure has been described by Fox (1965) and the following represents a summary of his recommendations:

(1) All standing trees on the site of the future plantation are poisoned with either sodium arsenite or *n*-butyl, 2,4,5-T.

(2) A creeping cover of mixed legumes is planted, but leaving a clean-weeded strip 6 ft wide along the planting rows, in which the young trees are set at 6–12 ft intervals, with rows 18–30 (usually 22) ft apart. The cover plants are progressively shaded out by the expanding crowns of the young rubber trees until, after 4–6 years, they are largely replaced by a sparse ground vegetation of shade-tolerant grasses and herbs.

(3) Quarterly rounds of foliage inspection begin at 1 year after planting. Diseased trees detected by foliage symptoms are removed and destroyed, and collar inspection (as by Napper's procedure, above) of neighbouring trees in the row is followed by the treatment detailed under (4) below; these trees are again collar-inspected and treated as necessary during subsequent rounds of foliage inspection.

(4) Infection sources thus discovered by collar inspection that lie *within* the clean-weeded planting row, together with infected sections of the young rubber roots, are extracted from the soil and

burned. Infection sources that lie outside the clean-weeded strip are left alone.

(5) But if an old stump outside the planting row *persistently* causes new infections, then it is isolated by a trench, all lateral roots are severed and infected ones removed, and the stump is left perched on its tap root.

Both the rationale of this revised control scheme, and the economies that have been effected as compared with the cost of Napper's procedure, are easily apparent, except for the special significance of the creeping leguminous cover plants as an aid to control of white root disease. Such a cover has, of course, a general agronomic value in shading the soil from the tropical sun and protecting it from erosion during heavy rainfall, in addition to its contribution of organic material and symbiotically fixed nitrogen to the soil. But over and above this, as was realized a number of years before Fox's recent account, these creeping leguminous cover plants seem specifically to reduce the incidence of white root disease in young rubber plantations. This effect of leguminous covers, first recorded by Napper, was originally unexpected and correspondingly difficult to understand, but a coherent and convincing explanation has now been advanced by Fox. First, the feeding roots of the cover plants, which are virtually immune to infection by *F. lignosus*, penetrate and thus disintegrate large infected roots left behind in the soil by the preceding stand of trees, and the smaller pieces into which these old roots are thus broken decay more quickly. Secondly, *F. lignosus* exhausts its food bases, provided by the old infected roots, more rapidly under the cover crop than under bare ground, because rhizomorphs are put out and grow along *any* root surfaces, whether susceptible to infection or not. Although older parts of the legume root system become infected by *F. lignosus*, the volume of an infected segment of legume root must often fail to generate an inoculum potential sufficient for infection of the rubber roots; moreover, such infected root sections of comparatively small volume become exhausted as food bases for *F. lignosus* long before the more massive infected roots left behind by the old trees would have lost their infectivity for rubber roots. Fox has therefore epitomized the mechanism of this cover-plant effect as 'the breakdown of large and potentially dangerous inocula into small and ineffective ones'.

More recently, Fox (1966) has described a novel control method of great promise, based upon the results of an experiment carried out by John (1958), also at the Rubber Research Institute of Malaya; this experiment demonstrated that if epiphytic advance of *F. lignosus* was arrested by removal of rhizomorphs at weekly intervals, then internal infection of the root was also halted. I have already explained the mechanism of this effect in Ch. 4 (p. 87). Fox therefore sought for a sufficiently persistent fungistatic preparation that would arrest the epiphytic advance of *F. lignosus* over the root surface; he found one in a formulation containing 15–20% pentachloronitrobenzene (PCNB). He recommends that this protectant be applied to the tap root in a band extending from a few inches above to a few inches below the zone of junction with the surface lateral roots, and also around the proximal 9 in of the laterals. The soil is then replaced and tamped down. The method is said to have given excellent results in control of *F. lignosus*, and is now recommended by the Rubber Research Institute for standard plantation practice. It is also notable as being the first to exploit in control practice our recent fuller comprehension of the mechanism of root infection by specialized, ectotrophic fungal pathogens.

This brief account of control measures practised against white root disease of rubber in Malaya would be incomplete without some reference to the possibility of basidiospore infection of exposed stump surfaces left by tree felling. In contrast with the behaviour of *Fomes annosus* basidiospores on coniferous stumps, as elucidated by Rishbeth, that of the basidiospores of *F. lignosus* is still open to much doubt. Fox (1969) has summarized evidence on this question from the Annual Reports of the Rubber Research Institute of Malaya, from 1959 onwards. Evidence that spore infection does sometimes occur has been provided by discovery of the mycelium of *F. lignosus* growing *downwards* from the exposed surface of a stump. On the other hand, numerous inoculation trials with basidiospores, carried out over a period of years and in a variety of ways, have met with only occasional success. Such

251

persistent failure suggests that the resistance of stump tissues after felling is too high to be overcome by the infectivity of most basidiospores. A similar conclusion can be drawn from Rishbeth's (1964) interim report on the results of numerous inoculations with basidiospores of *Armillaria mellea*; he obtained infection of hardwood stumps only in some 3 % of those inoculated. Certain treatments, probably injurious to the living wood tissues, promoted infection, and stumps showing minimal regrowth were more susceptible to basidiospore infection. Both observations suggest that host resistance of stump tissues of broad-leaved trees is more effective in preventing basidiospore infection by *A. mellea* than is that of coniferous stump tissues in excluding basidiospore infection by *F. annosus*. In making this comparison between two separate investigations, one often variable factor, i.e. the personality of the investigator, is fortunately common to both and can thus be eliminated.

The role of basidiospore infection in the epidemiology of some other tropical root diseases has been less difficult to assess. There has never been much doubt about the infectivity of the basidiospores produced by *Ustulina zonata* and by *Fomes noxius* Corner, respectively; both fungi cause root diseases of rubber in Malaya and also of tea in Ceylon, and their basidiospores readily infect exposed wood surfaces, even including those produced by branch pruning. Resistance of pruning wounds to infection is characteristic of that possessed by a standing tree in full vigour, and is thus considerably higher than that possessed by the tissues of a tree stump left by felling.

Other root-disease problems

The damage caused by any particular root-infecting pathogen can vary quite widely from one region or country to another; according to circumstances, the control scheme either has to be tightened up or it may be relaxed. For example, less stringent measures need to be taken in Ceylon than in Malaya against the white root disease of rubber. Thus in Ceylon, Peries, Fernando & Samaraweera (1963) have declared that the one essential measure that has to be taken during the early life of a young plantation is the tracing out and removal of the food-bases from which *Fomes lignosus* is causing infections. Results of a field experiment showed that if this were done, then no additional advantage accrued from the cutting out of infected sections of the rubber roots. This relaxation of control measures that is permissible in Ceylon, however, is partly due to the fact that rubber planters do not have to contend with another serious pathogen, *Ganoderma pseudoferreum* (Wakef.) van Over. & Steinm., which causes the equally feared red root disease in Malaya.

Again, a control measure that works well against a pathogen in one region or country may fail altogether in another. Thus the ring-barking control method instituted by R. Leach (1937, 1939) against *Armillaria mellea* in East Africa (Ch. 1, p. 8) has continued to work well there as a means for reducing the volume of infected roots left behind in the soil after clearing the land of forest or a preceding plantation crop (Wiehe, 1952). But results published by Redfern (1968) suggest that ring-barking before the felling of English woodland is likely to do nothing to reduce the incidence of Armillaria root disease in a young plantation established immediately after clearing of the site, and may even make it worse. In 1957 the English Forestry Commission set up an experiment to test the effects of ring-barking and of tree poisoning with 2,4,5-T, respectively, on subsequent invasion of the root systems by *A. mellea*. The trees were felled 2 years after these treatments and, 5 years after that, Redfern assessed the proportion of stump roots infected by *A. mellea* in each treatment-series. Amongst eighty roots taken from twenty stumps in each treatment-series, 90% were infected after ring-barking, 91% after tree poisoning, and 80% after no treatment (control). Redfern has ascribed the failure of these two treatments to reduce stump-root infection by *A. mellea* to the fact that this fungus is more widely and abundantly distributed under most English woodlands than under most East African forests. In England, moreover, *A. mellea* usually produces abundant rhizomorphs from its woody food bases lying in the soil; in East Africa it usually does not, and spread from one root to another requires contact between the two (Ch. 4, p. 102). It seems, therefore, that when *A. mellea* is so abundant in the soil as it is in most English woodlands, it can colonize the dying or dead roots ahead of saprophytic fungi; the more rapid decline in root resistance induced by ring-barking and by tree poisoning seems, according to Redfern's figures, actually to have benefitted *A. mellea*. This seems to be a good example of the *advantage of position* in saprophytic substrate colonization, which I discussed in Ch. 5 (p. 119). Other observations made by Redfern were consistent with his explanation. Stump roots from both ring-barked and poisoned trees were in a more advanced state of decay than were those taken from the control series; this indicated that they had become infected *earlier* by *A. mellea*. The more rotted roots taken from stumps in the two treated series produced a smaller crop of rhizomorphs, when tested, than did the less rotted roots from the control series. Redfern was thus able to conclude that the precocious loss of root resistance induced by both these treatments had accelerated both root invasion by *A. mellea* and its subsequent decline in vigour, due to approaching exhaustion of its substrate. The corollary of this conclusion was that the tree-killing treatments might prove useful in actual forestry practice for accelerating the cycle of *A. mellea* from first infection to substrate exhaustion; advantage of such a method could only be taken, however, when replanting of a woodland area could be delayed for a sufficient period of years after clear felling.

Apparently new root diseases have sometimes appeared on a widespread and devastating scale when a recently introduced species of crop plant has proved to be particularly susceptible to an existing pathogen. This is exemplified by the appearance of basal stem rot of oil palm (*Elaeis guineensis*) in Malaya, where there has been a spectacular increase in the acreage of this crop since 1950. Outbreaks of the disease have been studied by Turner (1965) and further particulars have been given by Turner & Bull (1967). The disease has broken out intensively where oil palms have been planted after coconut palms, especially on clay soils in coastal areas, and it is caused by certain species of *Ganoderma*. These fungi are commercially of minor importance as pathogens of the coconut palm but, as Turner has demonstrated, they are the dominant fungal colonizers of coconut stumps after felling, and also of the trunks if these are buried in the soil. If oil palms are to follow coconuts, then the only way of avoiding severe disease losses from these *Ganoderma* species is bodily removal of the old coconut palms from the site and their destruction by burning. Fortunately for the growing oil palm industry, the rising cost of manual labour in Malaya has been more than offset by the increase in efficiency of machinery available to planters. The old coconut palms are pushed over with a heavy bulldozer, and the thick mass of roots comes up with the base of the trunk; the operation must be carried out in dry weather, so that the trunks with their roots can be stacked and burned.

Despite Turner's successful solution of this particular problem, it is well for plant pathologists to remind themselves that solution of a disease-control problem by such a method may be a triumph for engineers but does not represent one for them; for plant pathologists it is merely the last resort in a desperate situation. The ultimate objective of plant pathology, like that of human medicine, is prevention, not cure. Resort to avoidable surgery on the body of the planting site is an admission of temporary defeat. It is better to employ crop residues to augment soil organic matter and fertility than to burn them; extraction of stump root systems damages the structure of the soil profile and this may sometimes aggravate the subsequent incidence of root disease. No one who has read my account of recent progress given in this book can doubt that these difficulties will be overcome; equally, however, no one can predict that other problems of at least comparable difficulty will not arise in the future. Most improvements in yield resulting from new crop husbandry practices bring new problems in their train; many of these problems will be problems for the plant pathologist.

Bibliography

Adams, J. E., Wilson, R. C., Hessler, L. E. & Ergle, D. R. (1939). Chemistry and growth of cotton in relation to soil fertility and root-rot. *Proc. Soil Sci. Soc. Am.* **4**, 329–332.

Ainsworth, G. C. (1961). *Ainsworth & Bisby's Dictionary of the Fungi*, 5th ed. Kew: Commonwealth Mycological Institute.

Aldrich, D. G. & Martin, J. P. (1952). Effect of fumigation on some chemical properties of soils. *Soil Sci.* **73**, 149–159.

Al-Shukri, M. M. (1969). The predisposition of the cotton plant to Verticillium and Fusarium wilt diseases by some major environmental factors. *J. Bot. un. Arab. Repub.* **12**, 13–25.

Altson, R. A. (1953). Diseases of the root system. *Rep. Rubb. Res. Inst. Malaya*, 1951, p. 34.

Armitage, P., Meynell, G. G. & Williams, T. (1965). Birth–death and other models for microbial infection. *Nature, Lond.* **207**, 570–572.

Armstrong, G. M. & Armstrong, Joanne K. (1948). Nonsusceptible hosts as carriers of wilt fusaria. *Phytopathology* **38**, 808–826.

Azevedo, Natalina F. S. (1963). Nitrogen utilization by four isolates of *Armillaria mellea*. *Trans. Br. mycol. Soc.* **46**, 281–284.

Bailey, I. W. & Vestal, Mary R. (1937). The significance of certain wood-destroying fungi in the study of the enzymatic hydrolysis of cellulose. *J. Arnold Arbor.* **18**, 196–205.

Baker, F. (1939). The disintegration of cellulose in the alimentary canal of herbivora. *Sci. Prog.* **134**, 287–301.

Baker, K. F. (1962). Principles of heat treatment of soil and planting material. *J. Aust. Inst. agric. Sci.* **28**, 118–126.

Baker, K. F., Flentje, N. T., Olsen, C. M. & Stretton, Helena M. (1967). Effect of antagonists on growth and survival of *Rhizoctonia solani* in soil. *Phytopathology* **57**, 591–597.

Baker, K. F. & Olsen, C. M. (1960). Aerated steam for soil treatment. *Phytopathology* **50**, 82.

Baker, K. F. & Snyder, W. C. (1965). *Ecology of Soil-borne Plant Pathogens. Prelude to Biological Control.* Berkeley: University of California Press.

Baker, R. (1965). The dynamics of inoculum. In *Ecology of Soil-borne Plant Pathogens.* K. F. Baker & W. C. Snyder ed., p. 419. Berkeley: University of California Press.

Baker, R. (1968). Mechanisms of biological control of soil-borne pathogens. *A. Rev. Phytopath.* **6**, 263–294.

Balis, C. & Kouyeas, V. (1968). Volatile inhibitors involved in soil mycostasis. *Annls Inst. phytopath. Benaki* **8**, 145–149.

Bateman, D. F. (1964). Cellulase and the Rhizoctonia disease of bean. *Phytopathology* **54**, 1372–1377.

Bateman, D. F. & Beer, S. V. (1965). Simultaneous production and synergistic action of oxalic acid and polygalacturonase during pathogenesis by *Sclerotium rolfsii*. *Phytopathology* **55**, 204–211.

Bateman, D. F. & Lumsden, R. D. (1965). Relation of calcium content and nature of the pectic substances in bean hypocotyls of different ages to susceptibility to an isolate of *Rhizoctonia solani*. *Phytopathology* **55**, 734–738.

Beckman, C. H. (1966). Cell irritability and localization of vascular infections in plants. *Phytopathology* **56**, 821–824.

Beckman, C. H. & Halmos, S. (1962). Relation of vascular occluding reactions in banana roots to pathogenicity of root-invading fungi. *Phytopathology* **52**, 893–897.

Beckman, C. H., Halmos, S. & Mace, M. E. (1962). The interaction of host, pathogen and soil temperature in relation to susceptibility to Fusarium wilt of bananas. *Phytopathology* **52**, 134–140.

Beckman, C. H. & Zaroogian, G. E. (1967). Origin and composition of vascular gel in infected banana roots. *Phytopathology* **57**, 11–13.

Bega, R. V. (1963). Root diseases of forest trees. *Fomes annosus*. *Phytopathology* **53**, 1120–1123.

Benzian, Blanche (1965). *Experiments on Nutrition Problems in Forest Nurseries*. Vols 1 and 2. London: H.M.S.O.

Bernard, N. (1909). L'evolution dans la symbiose. Les orchidées et leur champignons commenseux. *Annls Sci. nat. (Bot.)* **9**, 1–96.

Bisby, G. R. (1939). *Trichoderma viride* Pers. ex Fries, and notes on *Hypocrea*. *Trans. Br. mycol. Soc.* **23**, 149–168.

Blackhurst, Frances M. & Wood, R. K. S. (1963). Resistance of tomato plants to *Verticillium albo-atrum*. *Trans. Br. mycol. Soc.* **46**, 385–392.

Blackwell, Elizabeth (1943*a*). The life history of *Phytophthora cactorum* (Leb. & Cohn) Schroet. *Trans. Br. mycol. Soc.* **26**, 71–89.

Blackwell, Elizabeth (1943*b*). On germinating the oospores of *Phytophthora cactorum*. *Trans. Br. mycol. Soc.* **26**, 93–103.

Blair, I. D. (1943). Behaviour of the fungus *Rhizoctonia solani* Kühn in the soil. *Ann. appl. Biol.* **30**, 118–127.

Blank, L. M. (1944). Effect of nitrogen and phosphorus on the yield and root rot responses of early and late varieties of cotton. *J. Am. Soc. Agron.* **36**, 875–888.

Bliss, D. E. (1946). The relation of soil temperature to the development of Armillaria root rot. *Phytopathology* **36**, 302–318.

Bliss, D. E. (1951). The destruction of *Armillaria mellea* in citrus soils. *Phytopathology* **41**, 665–683.

Bochow, H. (1961). Beobachtungen über die spontane Keimung von *Plasmodiophora brassicae* Wor. *Z. Nach. PflSchutz* **15**, 53–55.

Bochow, H. & Seidel, D. (1964). Beiträge zur Frage des Einflusses einer organischen Dungüng auf den Befall von Pflanzen durch parasitische Pilze. IV. Wirkungen einer Stallmist-bzw. Strohdüngung auf *Plasmodiophora brassicae* Wor., *Ophiobolus graminis* Sacc. und *Helminthosporium sativum* P., K. & B. *Phytopath. Z.* **51**, 291–310.

Bollard, E. G. (1960). Transport in the xylem. *A. Rev. Pl. Physiol.* **11**, 141–166.

Börner, H. (1959). The apple replant problem. I. The excretion of phlorizin from apple root residues and its role in the soil sickness problem. *Contr. Boyce Thompson Inst. Pl. Res.* **20**, 39–56.

Börner, H. (1960). Liberation of organic substances from higher plants and their role in the soil sickness problem. *Bot. Rev.* **26**, 393–424.

Bosma, W. A. (1946). De resultaten van enkele proefnemingen. In: Verslag van den op 25 februari 1946 te Marknesse gehouden ontwikkelingdag voor de langboukundige opzichters. *Versl. Dir. Weiringermeer (Noordoostpolderweken)* no. 3, 3–19.

Bosma, W. A. (1962). Het verband tussen de rijping der gronden in de Ijsselmeerpolders en de verbouw van landbouwgewassen. *Van Zee Ld* **32**, 71–97.

Boyle, L. W. (1961). The ecology of *Sclerotium rolfsii*, with emphasis on the role of saprophytic media. *Phytopathology* **51**, 117–119.

Bremer, H. (1923). Untersuchungen über Biologie und Bekämpfung des Erregers der Kohlhernie, *Plasmodiophora brassicae* Woronin. *Landw. Jbr.* **59**, 227–243.

Bremer, H. (1924). Untersuchungen über Biologie und Bekämpfung des Erregers der Kohlhernie, *Plasmodiophora brassicae* Woronin. II. Kohlhernie und Bodenazidität. *Landw. Jbr.* **59**, 673–685.

Brian, P. W. (1960). Antagonistic and competitive mechanisms limiting survival and activity of fungi in soil. In *The Ecology of Soil Fungi*, D. Parkinson & J. S. Waid ed., pp. 115–129. Liverpool University Press.

Brian, P. W. (1967). The Leeuwenhoek Lecture, 1966. Obligate parasitism in fungi. *Proc. R. Soc., B* **168**, 101–118.

Brian, P. W., Elson, G. W. & Lowe, D. (1956). Production of patulin in apple fruits by *Penicillium expansum*. *Nature, Lond.* **178**, 263–264.

Brooks, D. H. (1965). Root infection by ascospores of *Ophiobolus graminis* as a factor in epidemiology of the take-all disease. *Trans. Br. mycol. Soc.* **48**, 237–248.

Brooks, D. H. & Dawson, M. G. (1968). Influence of direct-drilling of winter wheat on incidence of take-all and eyespot. *Ann. appl. Biol.* **61**, 57–64.

Brooks, F. T. (1908). Observations on the biology of *Botrytis cinerea*. *Ann. Bot.* **22**, 479–487.

Brooks, F. T. (1915). A disease of plantation rubber caused by *Ustulina zonata*. *New Phytol.* **14**, 152–164.

Brown, G. E. & Kennedy, B. W. (1966). Effect of oxygen concentration on Pythium seed rot of soybean. *Phytopathology* **56**, 407–411.

Brown, W. (1922*a*). Studies in the physiology of parasitism. VIII. On the exosmosis of substances from the host tissue into the infection drop. *Ann. Bot.* **36**, 101–119.

Brown, W. (1922*b*). On the germination and growth of fungi at various temperatures and in various concentrations of oxygen and of carbon dioxide. *Ann. Bot.* **36**, 257–283.

Brown, W. (1922c). Studies in the physiology of parasitism. IX. The effect on the germination of fungal spores of volatile substances arising from plant tissues. *Ann. Bot.* **36**, 285–300.

Brown, W. (1936). The physiology of host–parasite relations. *Bot. Rev.* **2**, 236–281.

Bruehl, G. W. & Lai, P. (1966). Prior colonization as a factor in the saprophytic survival of several fungi in wheat straw. *Phytopathology* **56**, 766–768.

Bruehl, G. W., Lai, P. & Huisman, O. (1964). Isolation of *Cephalosporium gramineum* from buried, naturally infested host debris. *Phytopathology* **54**, 1035–1036.

Burgeff, H. (1936). *Samenkeimung der Orchideen.* Jena: Gustaff Fischer.

Burges, A. (1939). Soil fungi and root infection. *Broteria* **8**, 64–81.

Burgess, L. W. & Griffin, D. M. (1967). Competitive saprophytic colonization of wheat straw. *Ann. appl. Biol.* **60**, 137–142.

Burgess, L. W. & Griffin, D. M. (1968). The relationship between the spore density of *Cochliobolus sativus* in soil and its saprophytic activity and parasitism. *Aust. J. exp. Agric. Anim. Husb.* **8**, 371–373.

Burke, D. W. (1965). Fusarium root rot of beans and behaviour of the pathogen in different soils. *Phytopathology* **55**, 1122–1126.

Burnet, F. M. (1953). *Natural History of Infectious Disease,* 2nd ed. Cambridge University Press.

Butler, E. J. (1907). An account of the genus *Pythium* and some chytridiaceae. *Mem. Dep. Agric. India, Bot. Ser.* **1**, no. 5.

Butler, F. C. (1953a). Saprophytic behaviour of some cereal root-rot fungi. I. Saprophytic colonization of wheat straw. *Ann. appl. Biol.* **40**, 284–297.

Butler, F. C. (1953b). Saprophytic behaviour of some cereal root-rot fungi. II. Factors influencing saprophytic colonization of wheat straw. *Ann. appl. Biol.* **40**, 298–304.

Butler, F. C. (1953c). Saprophytic behaviour of some cereal root-rot fungi. III. Saprophytic survival in wheat straw buried in soil. *Ann. appl. Biol.* **40**, 305–311.

Butler, F. C. (1959). Saprophytic behaviour of some cereal root-rot fungi. IV. Saprophytic survival in soils of high and low fertility. *Ann. appl. Biol.* **47**, 28–36.

Butler, Gillian M. (1957). The development and behaviour of mycelial strands in *Merulius lacrymans* (Wulf.) Fr. I. Strand development during growth from a food-base through a non-nutrient medium. *Ann. Bot.* **21**, 523–537.

Butler, Gillian M. (1958). The development and behaviour of mycelial strands in *Merulius lacrymans* (Wulf.) Fr. II. Hyphal behaviour during strand formation. *Ann. Bot.* **22**, 219–236.

Buxton, E. W. (1957). Some effects of pea root exudates on physiologic races of *Fusarium oxysporum* Fr. f. *pisi* (Lindford) Snyder & Hansen. *Trans. Br. mycol. Soc.* **40**, 145–154.

Buxton, E. W. (1962). Root exudates from banana and their relationship to strains of the *Fusarium* causing Panama wilt. *Ann. appl. Biol.* **50**, 269–282.

Byther, R. (1965). Ecology of plant pathogens in soil. V. Inorganic nitrogen utilization as a factor of competitive saprophytic ability of *Fusarium roseum* and *F. solani*. *Phytopathology* **55**, 852–858.

Campbell, A. H. (1933). Zone lines in plant tissues. I. The black lines formed by *Xylaria polymorpha* (Pers.) Grev. in hardwoods. *Ann. appl. Biol.* **20**, 123–145.

Campbell, A. H. (1934). Zone lines in plant tissues. II. The black lines formed by *Armillaria mellea* (Vahl) Quél. *Ann. appl. Biol.* **21**, 1–22.

Chacko, C. I. & Lockwood, J. L. (1966). A quantitative method for assaying soil fungistasis. *Phytopathology* **56**, 576–577.

Chambers, S. C. & Flentje, N. T. (1969). Relative effects of soil nitrogen and soil organisms on survival of *Ophiobolus graminis*. *Aust. J. biol. Sci.* **22**, 275–278.

Chang, Yung (1967). The fungi of wheat straw compost. II. Biochemical and physiological studies. *Trans. Br. mycol. Soc.* **50**, 667–677.

Chesters, C. G. C. (1940). A method of isolating soil fungi. *Trans. Br. mycol. Soc.* **24**, 352–355.

Chet, I. & Henis, Y. (1968). The control mechanism of sclerotial formation in *Sclerotium rolfsii* Sacc. *J. gen. Microbiol.* **54**, 231–236.

Chet, I., Henis, Y. & Mitchell, R. (1966). The morphogenetic effect of sulphur-containing amino acids, glutathione and iodoacetic acid on *Sclerotium rolfsii* Sacc. *J. gen. Microbiol.* **45**, 541–546.

Chi, C. C. & Hanson, E. W. (1962). Interrelated effects of environment and age of alfalfa and red clover seedlings on susceptibility to *Pythium de Baryanum*. *Phytopathology* **52**, 985–989.

Chinn, S. H. F. (1953). A slide technique for the study of fungi and actinomycetes in soil with special reference to *Helminthosporium sativum*. *Can. J. Bot.* **31**, 718–724.

Chinn, S. H. F. (1967). Differences in fungistasis in some Saskatchewan soils with special reference to *Cochliobolus sativus*. *Phytopathology* **57**, 224–226.

Chinn, S. H. F. & Tinline, R. D. (1963). Spore germinability in soil as an inherent character of *Cochliobolus sativus*. *Phytopathology* **53**, 1109–1112.

Chinn, S. H. F. & Tinline, R. D. (1964). Inherent germinability and survival of spores of *Cochliobolus sativus*. *Phytopathology* **54**, 349–352.

Christou, T. (1962). Penetration and host–parasite relationships of *Rhizoctonia solani* in the bean plant. *Phytopathology* **52**, 381–389.

Clark, F. E. (1942). Experiments towards the control of the take-all disease of wheat and the Phymatotrichum root rot of cotton. *Tech. Bull. U.S. Dep. Agric.* no. 835.

Cobb, F. W., Krstic, M., Zavarin, E. & Barber, H. W. (1968). Inhibitory effects of volatile oleoresin components on *Fomes annosus* and four *Ceratocystis* species. *Phytopathology* **58**, 1327–1335.

Cochrane, Jean C., Cochrane, V. W., Simon, F. G. & Spaeth, J. (1963). Spore germination and carbon metabolism in *Fusarium solani*. I. Requirements for spore germination. *Phytopathology* **53**, 1155–1160.

Cochrane, V. W. (1948). The role of plant residues in the etiology of root rot. *Phytopathology* **38**, 185–196.

Cochrane, V. W. (1949). Crop residues as causative agents of root rots of vegetables. *Bull. Conn. agric. Exp. Stn* no. 526.

Cochrane, V. W. (1958). *Physiology of Fungi*. New York: Wiley.

Cochrane, V. W. (1966). Respiration and spore germination. In *The Fungus Spore*, Colston Papers no. 18, M. F. Madelin ed., pp. 201–215. London: Butterworth.

Coley-Smith, J. R. (1959). Studies of the biology of *Sclerotium cepivorum* Berk. II. Host range; persistence and viability of sclerotia. *Ann. appl. Biol.* **47**, 511–518.

Coley-Smith, J. R. (1960). Studies of the biology of *Sclerotium cepivorum* Berk. IV. Germination of sclerotia. *Ann. appl. Biol.* **48**, 8–18.

Coley-Smith, J. R., Dickinson, D. J., King, J. E. & Holt, R. W. (1968). The effect of species of *Allium* on soil bacteria in relation to germination of sclerotia of *Sclerotium cepivorum* Berk. *Ann. appl. Biol.* **62**, 103–111.

Coley-Smith, J. R. & Holt, R. W. (1966). The effect of species of *Allium* on germination in soil of sclerotia of *Sclerotium cepivorum* Berk. *Ann. appl. Biol.* **58**, 273–278.

Coley-Smith, J. R., King, J. E., Dickinson, D. J. & Holt, R. W. (1967). Germination of sclerotia of *Sclerotium cepivorum* Berk. under aseptic conditions. *Ann. appl. Biol.* **60**, 109–115.

Colhoun, J. (1953). A study of the epidemiology of clubroot disease of Brassicae. *Ann. appl. Biol.* **40**, 262–283.

Colhoun, J. (1961). Spore load, light intensity and plant nutrition as factors influencing the incidence of clubroot of Brassicae. *Trans. Br. mycol. Soc.* **44**, 593–600.

Cook, R. J. (1968). Fusarium root and foot rot of cereals in the Pacific Northwest. *Phytopathology* **58**, 127–131.

Cook, R. J. & Bruehl, G. W. (1968). Relative significance of parasitism versus saprophytism in colonization of wheat straw by *Fusarium roseum* 'Culmorum' in the field. *Phytopathology* **58**, 306–308.

Cook, R. J. & Flentje, N. T. (1967). Chlamydospore germination and germling survival of *Fusarium solani* f. *pisi* in soil as affected by soil water and pea seed exudation. *Phytopathology* **57**, 178–182.

Cook, R. J. & Schroth, M. N. (1965). Carbon and nitrogen compounds and germination of chlamydospores of *Fusarium solani* f. *phaseoli*. *Phytopathology* **55**, 254–256.

Cook, R. J. & Snyder, W. C. (1965). Influence of host exudates on growth and survival of germlings of *Fusarium solani* f. *phaseoli* in soil. *Phytopathology* **55**, 1021–1025.

Cook, W. R. I. & Schwartz, E. J. (1930). The life-history, cytology and method of infection of *Plasmodiophora brassicae* Woron., the cause of finger-and-toe disease of cabbages and other crucifers. *Phil. Trans. R. Soc., B,* **218**, 283–314.

Cowling, E. B. (1963). Structural features of cellulose that influence its susceptibility to enzymatic hydrolysis. In *Enzymic Hydrolysis of Cellulose and Related Materials*, E. T. Reese ed. Oxford: Pergamon Press.

Cowling, E. B. & Merrill W. (1966). Nitrogen in wood and its role in wood deterioration. *Can. J. Bot.* **44**, 1539–1554.

Cruickshank, I. A. M. (1962). Phytoalexins. *A. Rev. Phytopath.* **1**, 351–374.

Cunningham, J. L. & Hagedorn, D. J. (1962). Attraction of *Aphanomyces euteiches* zoospores to pea and other plant roots. *Phytopathology* **52**, 616–618.

Dade, H. A. (1927). 'Collar Crack' of cacao, *Armillaria mellea* (Vahl) Fr. *Bull. Dep. Agric. Gold Cst* no. 5.

Davies, F. R. (1935). Superiority of silver nitrate over mercuric chloride for surface sterilization in the isolation of *Ophiobolus graminis* Sacc. *Can. J. Res., C* **13**, 168–173.

Dawson, J. R., Johnson, R. A. H., Adams, P. & Last, F. T. (1965). Influence of steam/air mixtures, when used for heating soil, on biological and chemical properties that affect seedling growth. *Ann. appl. Biol.* **56**, 243–251.

Dawson, J. R., Kilby, A. A. T., Ebben, Marion H. & Last, F. T. (1967). The use of steam/air mixtures for partially sterilizing soils infested with cucumber root rot pathogens. *Ann. appl. Biol.* **60**, 215–222.

Day, Sarah C. (1969). The morphogenesis of mycelial strands in the timber dry rot fungus, *Merulius lacrymans* (Wulf.) Fr. Ph.D. thesis, University of Cambridge.

Deverall, B. J. & Wood, R. K. S. (1961). Infection of bean plants (*Vicia faba* L.) with *Botrytis cinerea* and *B. fabae*. *Ann. appl. Biol.* **49**, 461–472.

Dickens, L. E. (1964). Eyespot footrot of winter wheat caused by *Cercosporella herpotrichoides*. *Mem. Cornell agric. Exp. Stn* no. 390.

Dickson, J. G. (1923). Influence of soil temperature and moisture on the development of the seedling blight of wheat and corn caused by *Gibberella saubinetii*. *J. agric. Res.* **23**, 837–869.

Dickson, J. G., Eckerson, S. H. & Link, K. P. (1923). The nature of resistance to seedling blight of cereals. *Proc. natn. Acad. Sci. U.S.A.* **9**, 434–439.

Dillon Weston, W. A. R. & Garrett, S. D. (1943). *Rhizoctonia solani* associated with a root rot of cereals in Norfolk. *Ann. appl. Biol.* **30**, 79.

Dimond, A. E. & Waggoner, P. E. (1953). On the nature and role of vivotoxins in plant disease. *Phytopathology* **43**, 229–235.

Dix, N. J. (1964). Colonization and decay of bean roots. *Trans. Br. mycol. Soc.* **47**, 285–292.

Dix, N. J. (1967). Mycostasis and root exudation: factors influencing the colonization of bean roots by fungi. *Trans. Br. mycol. Soc.* **50**, 23–31.

Dobbs, C. G. & Gash, M. J. (1965). Microbial and residual mycostasis in soils. *Nature, Lond.* **207**, 1354–1356.

Dobbs, C. G. & Hinson, W. H. (1953). A widespread fungistasis in soils. *Nature, Lond.* **172**, 197.

Dodman, R. L., Barker, K. R. & Walker, J. C. (1968a). Modes of penetration by different isolates of *Rhizoctonia solani*. *Phytopathology* **58**, 31–33.

Dodman, R. L., Barker, K. R. & Walker, J. C. (1968b). A detailed study of the different modes of penetration by *Rhizoctonia solani*. *Phytopathology* **58**, 1271–1276.

Downie, D. G. (1957). *Corticium solani*—an orchid endophyte. *Nature, Lond.* **179**, 160.

Downie, D. G. (1959a). *Rhizoctonia solani* and orchid seed. *Trans. bot. Soc. Edinb.* **37**, 279–285.

Downie, D. G. (1959b). The mycorrhiza of *Orchis purpurella*. *Trans. bot. Soc. Edinb.* **38**, 16–29.

Driver, C. H. (1963). Further data on borax as a control of surface infection of slash pine stumps by *Fomes annosus*. *Pl. Dis. Reptr* **47**, 1006–1009.

Duggar, B. M. (1915). *Rhizoctonia crocorum* (Pers.) DC. and *R. solani* Kühn (*Corticium vagum* B. & C.) with notes on other species. *Ann. Mo. bot. Gdn* **2**, 403–458.

Dukes, P. D. & Apple, J. L. (1961). Chemotaxis of zoospores of *Phytophthora parasitica* var. *nicotianae* by plant roots and certain chemical solutions. *Phytopathology* **51**, 195–197.

Dunlap, A. A. (1943). Inhibition of Phymatotrichum sclerotia formation by sulphur autoclaved with soil. *Phytopathology* **33**, 1205–1208.

Dwivedi, R. S. & Garrett, S. D. (1968). Fungal competition in agar plate colonization from soil inocula. *Trans. Br. mycol. Soc.* **51**, 95–101.

Ergle, D. R. (1948). The carbohydrate metabolism of germinating Phymatotrichum sclerotia with special reference to glycogen. *Phytopathology* **38**, 142–151.

Falck, R. (1912). Die Merulius Fäule des Bauholzes. *Hausschwammforsch.* **6**, 1–405.

Faris, J. A. (1921). Violet root rot (*Rhizoctonia crocorum* DC.) in the United States. *Phytopathology* **11**, 412–423.

Fedorintchik, N. S. (1935). *Rev. appl. Mycol.* **15**, 624.

Fellows, H. (1941). Effect of certain environmental conditions on the prevalence of *Ophiobolus graminis* in the soil. *J. agric. Res.* **63**, 715–726.

Fellows, H. & Ficke, C. H. (1934). Cereal and forage crop disease investigations. *Rep. Kans. agric. Exp. Stn*, 1932–4, pp. 94–97.

Findlay, W. P. K. (1951). The development of *Armillaria mellea* rhizomorphs in a water tunnel. *Trans. Br. mycol. Soc.* **34**, 146.

Flentje, N. T. (1957). Studies on *Pellicularia filamentosa* (Pat.) Rogers.

III. Host penetration and resistance, and strain specialization. *Trans. Br. mycol. Soc.* **40**, 322–336.

Flentje, N. T. (1969). Genetical aspects of pathogenic and saprophytic behaviour in soil-borne fungi. Basidiomycetes with special reference to *Thanatephorus cucumeris*. In *Root Diseases and Soil-borne Pathogens*, T. A. Toussoun, R. V. Bega & P. E. Nelson ed. (In Press.) Berkeley: University of California Press.

Flentje, N. T., Dodman, R. L. & Kerr, A. (1963). The mechanism of host penetration by *Thanatephorus cucumeris*. *Aust. J. biol. Sci.* **16**, 784–799.

Foster, J. W. (1949). *Chemical Activities of Fungi*. New York: Academic Press.

Foster, R. E. & Walker, J. C. (1947). Predisposition of tomato to Fusarium wilt. *J. agric. Res.* **74**, 165–185.

Fox, R. A. (1965). The role of biological eradication in root-disease control in replantings of *Hevea brasiliensis*. In *Ecology of Soil-borne Plant Pathogens*, K. F. Baker & W. C. Snyder ed., pp. 348–362. Berkeley: University of California Press.

Fox, R. A. (1966). White root disease of *Hevea brasiliensis*: collar protectant dressings. *J. Rubb. Res. Inst. Malaya* **19**, 231–241.

Fox, R. A. (1969). A comparison of methods of dispersal, survival and parasitism in some fungi causing root diseases of tropical plantation crops. In *Root Diseases and Soil-borne Pathogens*, T. A. Toussoun, R. V. Bega & P. E. Nelson ed. (In Press.) Berkeley: University of California Press.

Fraser, Lilian (1942). Phytophthora root rot of citrus. *J. Aust. Inst. agric. Sci.* **8**, 101–105.

Fraser, Lilian (1949). A gummosis disease of citrus in relation to its environment. *Proc. Linn. Soc. N.S.W.* **74**, 5–18.

Fraymouth, Joan (1956). Haustoria of the Peronosporales. *Trans. Br. mycol. Soc.* **39**, 79–107.

Garraway, M. O. & Weinhold, A. R. (1968). Period of access to ethanol in relation to carbon utilization and rhizomorph initiation and growth in *Armillaria mellea*. *Phytopathology* **58**, 1190–1191.

Garren, K. H. (1961). Control of *Sclerotium rolfsii* through cultural practices. *Phytopathology* **51**, 120–124.

Garren, K. H. (1964). Inoculum potential and differences among peanuts in susceptibility to *Sclerotium rolfsii*. *Phytopathology* **54**, 279–281.

Garrett, S. D. (1934). Factors affecting the severity of take-all. I. The importance of micro-organisms. *J. Dep. Agric. S. Aust.* **37**, 664–674.

Garrett, S. D. (1936). Soil conditions and the take-all disease of wheat. *Ann. appl. Biol.* **23**, 667–699.

Garrett, S. D. (1937). Soil conditions and the take-all disease of wheat. II. The relation between soil reaction and soil aeration. *Ann. appl. Biol.* **24**, 747–751.

Garrett, S. D. (1938). Soil conditions and the take-all disease of wheat. III. Decomposition of the resting mycelium of *Ophiobolus graminis* in infected wheat stubble buried in the soil. *Ann. appl. Biol.* **25**, 742–766.

Garrett, S. D. (1939). Soil conditions and the take-all disease of wheat. IV. Factors limiting infection by ascospores of *Ophiobolus graminis*. *Ann. appl. Biol.* **26**, 47–55.

Garrett, S. D. (1940). Soil conditions and the take-all disease of wheat. V. Further experiments on the survival of *Ophiobolus graminis* in infected wheat stubble buried in the soil. *Ann. appl. Biol.* **27**, 199–204.

Garrett, S. D. (1941). Soil conditions and the take-all disease of wheat. VI. The effect of plant nutrition upon disease resistance. *Ann. appl. Biol.* **28**, 14–18.

Garrett, S. D. (1944). Soil conditions and the take-all disease of wheat. VIII. Further experiments on the survival of *Ophiobolus graminis* in infected wheat stubble. *Ann. appl. Biol.* **31**, 186–191.

Garrett, S. D. (1946). A study of violet root rot. Factors affecting production and growth of mycelial strands in *Helicobasidium purpureum* Pat. *Trans. Br. mycol. Soc.* **29**, 114–127.

Garrett, S. D. (1948). Soil conditions and the take-all disease of wheat. IX. Interaction between host plant nutrition, disease escape and disease resistance. *Ann. appl. Biol.* **35**, 14–17.

Garrett, S. D. (1949). A study of violet root rot. II. Effect of substratum on survival of *Helicobasidium purpureum* colonies in the soil. *Trans. Br. mycol. Soc.* **32**, 217–223.

Garrett, S. D. (1950). Ecology of the root-inhabiting fungi. *Biol. Rev.* **25**, 220–254.

Garrett, S. D. (1951). Ecological groups of soil fungi: a survey of substrate relationships. *New Phytol.* **50**, 149–166.

Garrett, S. D. (1953). Rhizomorph behaviour in *Armillaria mellea* (Vahl) Quél. I. Factors controlling rhizomorph initiation by *A. mellea* in pure culture. *Ann. Bot.* **17**, 63–79.

Garrett, S. D. (1956a). *Biology of Root-infecting Fungi*. Cambridge University Press.

Garrett, S. D. (1956b). Rhizomorph behaviour in *Armillaria mellea* (Vahl) Quél. II. Logistics of infection. *Ann. Bot.* **20**, 193–209.

Garrett, S. D. (1957). Effect of a soil microflora selected by carbon disulphide fumigation on survival of *Armillaria mellea* in woody host tissues. *Can. J. Microbiol.* **3**, 135–149.

Garrett, S. D. (1958). Inoculum potential as a factor limiting lethal action by *Trichoderma viride* Fr. on *Armillaria mellea* (Fr.) Quél. *Trans. Br. mycol. Soc.* **41**, 157–164.

Garrett, S. D. (1960a). Inoculum potential. In *Plant Pathology*, J. G. Horsfall & A. E. Dimond ed., vol. 3, pp. 23–56. New York: Academic Press.

Garrett, S. D. (1960b). Rhizomorph behaviour in *Armillaria mellea* (Fr.) Quél. III. Saprophytic colonization of woody substrates in soil. *Ann. Bot.* **24**, 275–285.

Garrett, S. D. (1962). Decomposition of cellulose in soil by *Rhizoctonia solani*. *Trans. Br. mycol. Soc.* **45**, 115–120.

Garrett, S. D. (1963). *Soil Fungi and Soil Fertility*. Oxford: Pergamon Press.

Garrett, S. D. (1965). Toward biological control of soil-borne plant pathogens. In *Ecology of Soil-borne Plant Pathogens*, K. F. Baker & W. C. Snyder ed., pp. 4–17. Berkeley: University of California Press.

Garrett, S. D. (1966a). Spores as propagules of disease. In *The Fungus Spore*, Colston Papers no. 18, M. F. Madelin ed., pp. 309–319. London: Butterworth.

Garrett, S. D. (1966b). Cellulose-decomposing ability of some cereal foot-rot fungi in relation to their saprophytic survival. *Trans. Br. mycol. Soc.* **49**, 57–68.

Garrett, S. D. (1967). Effect of nitrogen level on survival of *Ophiobolus graminis* in pure culture on cellulose. *Trans. Br. mycol. Soc.* **50**, 519–524.

Garrett, S. D. & Buddin, W. (1947). Control of take-all under the Chamberlain system of intensive barley growing. *Agriculture, Lond.* **54**, 425–426.

Garrett, S. D. & Mann, H. H. (1948). Soil conditions and the take-all disease of wheat. X. Control of the disease under continuous cultivation of a spring-sown cereal. *Ann. appl. Biol.* **35**, 435–442.

Gericke, W. F. (1940). *The Complete Guide to Soilless Gardening*. London: Putnam.

Gerlagh, M. (1966). A perspex box for microscopic observation of living roots. *Neth. J. Pl. Path.* **72**, 248–249.

Gerlagh, M. (1968). Introduction of *Ophiobolus graminis* into new polders and its decline. *Meded. Lab. Phytopath. Wageningen* no. 241.

Gibbs, J. G. (1939). Factors influencing the control of clubroot (*Plasmodiophora brassicae*). *N.Z. Jl Sci. Technol.*, A **20**, 409–412.

Gibbs, J. N. (1967). A study of the epiphytic growth habit of *Fomes annosus*. *Ann. Bot.* **31**, 755–774.

Gibbs, J. N. (1968). Resin and the resistance of conifers to *Fomes annosus*. *Ann. Bot.* **32**, 649–665.

Gibson, I. A. S. (1961). A note on variation between isolates of *Armillaria mellea* (Vahl ex Fr.) Kummer. *Trans. Br. mycol. Soc.* **44**, 123–128.

Gibson, I. A. S. & Goodchild, N. A. (1960). *Armillaria mellea* in Kenya forests. *E. Afr. agric. For. J.* **26**, 142–143.

Glynne, Mary D. (1953). Production of spores by *Cercosporella herpotrichoides*. *Trans. Br. mycol. Soc.* **36**, 46–51.

Glynne, Mary D. (1965). Crop sequence in relation to soil-borne pathogens. In *Ecology of Soil-borne Plant Pathogens*, K. F. Baker & W. C. Snyder ed., pp. 423–435. Berkeley: University of California Press.

Gordee, R. S. & Porter, C. L. (1961). Structure, germination and physiology of microsclerotia of *Verticillium albo-atrum*. *Mycologia* **53**, 171–182.

Goring, C. A. I. (1967). Physical aspects of soil in relation to the action of soil fungicides. *A. Rev. Phytopath.* **5**, 285–318.

Goring, C. A. I. (1969). Physical soil factors and soil fumigant action. In *Root Diseases and Soil-borne Pathogens*, T. A. Toussoun, R. V. Bega & P. E. Nelson ed. (In Press.) Berkeley: University of California Press.

Gottlieb, D. (1950). The physiology of spore germination in fungi. *Bot. Rev.* **16**, 229–257.

Gottlieb, D. (1966). Biosynthetic processes in germinating spores. In *The Fungus Spore*, Colston Papers no. 18, M. F. Madelin ed., pp. 217–234. London: Butterworth.

Grainger, J. (1959). Disease control through intimate mixing of mercuric oxide with soil. *Phytopathology* **49**, 627–633.

Gras, N. S. B. (1940). *A History of Agriculture*, 2nd ed. New York: Crofts.

Greenwood, D. J. & Nye, P. H. (1968). Carbon dioxide distribution in soil. *Rep. natn. Veg. Res. Stn*, Wellesbourne, 1967, pp. 34–35.

Gregory, P. H. (1961). *The Microbiology of the Atmosphere.* London: Leonard Hill.

Griffin, D. M. (1963a). Soil moisture and the ecology of soil fungi. *Biol. Rev.* **38**, 141–166.

Griffin, D. M. (1963b). Soil physical factors and the ecology of fungi. II. Behaviour of *Pythium ultimum* at small soil water suctions. *Trans. Br. mycol. Soc.* **46**, 368–372.

Griffin, D. M. (1968). Observations on fungi growing in a translucent particulate matrix. *Trans. Br. mycol. Soc.* **51**, 319–322.

Griffin, D. M. (1969). Effect of soil moisture and aeration on fungal activity—an introduction. In *Root Diseases and Soil-borne Pathogens*, T. A. Toussoun, R. V. Bega & P. E. Nelson ed. (In Press.) Berkeley: University of California Press.

Griffiths, R. L. (1933). Take-all. Incidence and control on the lighter soils of the mallee. *J. Dep. Agric. S. Aust.* **36**, 774–778.

Grogan, R. G. & Campbell, R. N. (1966). Fungi as vectors and hosts of viruses. *A. Rev. Phytopath.* **4**, 29–52.

Halvorsen, H. O. (1935). The effect of chance on the mortality of experimentally infected animals. *J. Bact.* **30**, 330–331.

Hansford, C. G. (1926). The Fusaria of Jamaica. *Kew Bull.* **7**, 257–288.

Harley, J. L. (1948). Mycorrhiza and soil ecology. *Biol. Rev.* **23**, 127–158.

Harley, J. L. (1968). Presidential Address. Fungal symbiosis. *Trans. Br. mycol. Soc.* **51**, 1–11.

Harley, J. L. (1969). *The Biology of Mycorrhiza*, 2nd ed. London: Leonard Hill.

Hartig, R. (1873). Vorläufige Mitteilung über den Parasitismus von *Agaricus melleus* und dessen Rhizomorphen. *Bot. Ztg* **31**, 295–297.

Heald, F. D. (1921). The relation of the spore-load to the percentage smut appearing in the crop. *Phytopathology* **11**, 269–278.

Heale, J. B. & Isaac, I. (1963). Wilt of lucerne caused by species of *Verticillium*. IV. Pathogenicity of *V. albo-atrum* and *V. dahliae* to lucerne and other crops; spread and survival of *V. albo-atrum* in soil and in weeds; effect upon lucerne production. *Ann. appl. Biol.* **52**, 439–451.

Heinze, P. M. & Andrus, C. F. (1945). Apparent localization of Fusarium wilt resistance in the Pan America tomato. *Am. J. Bot.* **32**, 62–66.

Hendrix, F. F. & Nielsen, L. W. (1958). Invasion and infection of crops other than the forma suscept by *Fusarium oxysporum* f. *batatas* and other formae. *Phytopathology* **48**, 224–228.

Hering, T. F. (1962). Infection cushions of *Helicobasidium purpureum* Pat. *Trans. Br. mycol. Soc.* **45**, 46–54.

Hiltner, L. (1904). Über neuere Erfahrungen und Probleme auf dem Gebiet der Bodenbakteriologie und unter besonderer Berücksichtigung der Gründüngung und Brache. *Arb. dtsch. Landw-Ges.* **98**, 59–78.

Hooker, W. J., Walker, J. C. & Link, K. P. (1945). Effects of two mustard oils on *Plasmodiophora brassicae* and their relation to resistance to clubroot. *J. agric. Res.* **70**, 63–78.

Hopp, H. (1938). The formation of coloured zones by wood-destroying fungi in culture. *Phytopathology* **28**, 601–620.

Hornby, D. (1969). Gravimetrical and mycological investigations of soil suspensions in the soil dilution plate technique. *J. appl. Bact.* **32**, 244–258.

Horton, J. C. & Keen, N. T. (1966). Sugar repression of endopolygalacturonase and cellulase synthesis during pathogenesis by *Pyrenochaeta terrestris* as a resistance mechanism in onion pink root. *Phytopathology* **56**, 908–916.

Hudson, H. J. (1968). The ecology of fungi on plant remains above the soil. *New Phytol.* **67**, 837–874.

Husain, S. S. & McKeen, W. E. (1963). Interactions between strawberry roots and *Rhizoctonia fragariae*. *Phytopathology* **53**, 541–545.

Hynes, H. J. (1937). Studies on Rhizoctonia root-rot of wheat and oats. *Sci. Bull. Dep. Agric. N.S.W.* no. 58.

Isaac, I. (1946). Verticillium wilt of sanfoin. *Ann. appl. Biol.* **33**, 28–34.

Isaac, I. (1949). A comparative study of pathogenic isolates of *Verticillium*. *Trans. Br. mycol. Soc.* **32**, 137–157.

Isaac, I. (1953). The spread of diseases caused by species of *Verticillium*. *Ann. appl. Biol.* **40**, 630–638.

Isaac, I. (1957). Wilt of lucerne caused by species of *Verticillium*. *Ann. appl. Biol.* **45**, 550–558.

Isaac, I. & MacGarvie, Q. D. (1962). Germination of resting bodies in *Verticillium* species. *Nature, Lond.* **195**, 826–827.

Jackson, R. M. (1958*a*). An investigation of fungistasis in Nigerian soils. *J. gen. Microbiol.* **18**, 248–258.

Jackson, R. M. (1958*b*). Some aspects of soil fungistasis. *J. gen. Microbiol.* **19**, 390–401.

Jalaluddin, M. (1967*a*). Studies on *Rhizina undulata*. I. Mycelial growth and ascospore germination. *Trans. Br. mycol. Soc.* **50**, 449–459.

Jalaluddin, M. (1967*b*). Studies on *Rhizina undulata*. II. Observations and experiments in East Anglian plantations. *Trans. Br. mycol. Soc.* **50**, 461–472.

John, K. P. (1958). Inoculation experiments with *Fomes lignosus* Klotzsch. *J. Rubb. Res. Inst. Malaya* **15**, 223–230.

Johnson, H., Holland, A. H., Paulus, A. O. & Wilhelm, S. (1962). Soil fumigation found essential for maximum strawberry yields in southern California. *Calif. Agric.* **16**, no. 10.

Johnson, J. (1919). The influence of heated soils on seed germination and plant growth. *Soil Sci.* **7**, 1–103.

Jong, W. H. de (1933). Het parasitisme van *Rigidoporus microporus* = *Fomes lignosus*, bij *Hevea brasiliensis*. *Archf. Rubbercult. Ned.-Indië* **17**, 83–104.

Jordan, H. V., Nelson, H. A. & Adams, J. E. (1939). Relation of fertilizers, crop residues and tillage to yields of cotton and incidence of root-rot. *Proc. Soil Sci. Soc. Am.* **4**, 325–328.

Kerr, A. (1956). Some interactions between plant roots and pathogenic soil fungi. *Aust. J. biol. Sci.* **9**, 45–52.

Kerr, A. (1964). The influence of soil moisture on infection of peas by *Pythium ultimum*. *Aust. J. biol. Sci.* **17**, 676–685.

Kerr, A. & Flentje, N. T. (1957). Host infection in *Pellicularia filamentosa* controlled by chemical stimuli. *Nature, Lond.* **179**, 204–205.

Kessler, K. J. (1966). Xylem sap as a growth medium for four tree wilt fungi. *Phytopathology* **56**, 1165–1169.

Keyworth, W. G. (1953). Verticillium wilt of the hop. VI. The relative roles of root and stem in the determination of wilt severity. *Ann. appl. Biol.* **40**, 344–361.

King, C. J. (1937). A method for the control of cotton root rot in the irrigated southwest. *Circ. U.S. Dep. Agric.* no. 425.

King, C. J. & Eaton, E. D. (1934). Influence of soil moisture on the longevity of cotton root rot sclerotia. *J. agric. Res.* **49**, 793–798.

King, C. J. & Loomis, H. F. (1929). Further studies of cotton root rot in Arizona with a description of a sclerotium stage of the fungus. *J. agric. Res.* **39**, 641–676.

King, J. E. & Coley-Smith, J. R. (1968). Effects of volatile products of *Allium* species and their extracts on germination of sclerotia of *Sclerotium cepivorum* Berk. *Ann. appl. Biol.* **61**, 407–414.

Ko, Wen-Hsiung & Lockwood, J. L. (1967). Soil fungistasis: relation to fungal spore nutrition. *Phytopathology* **57**, 894–901.

Kole, A. P. & Gielink, A. J. (1962). Electron microscope observations on the resting-spore germination of *Plasmodiophora brassicae*. *Proc. K. ned. Akad. Wet.*, C **65**, 117–121.

Kole, A. P. & Philipsen, P. J. J. (1956). Over de vatbaarheid van niet-kruisbloemige planten voor het zoosporangium-stadium van *Plasmodiophora brassicae* Woron. *Tijdschr. PlZiekt.* **62**, 167–170.

Kommedahl, T. (1966). Relation of exudates of pea roots to germination of spores in races of *Fusarium oxysporum* f. *pisi*. *Phytopathology* **56**, 721–722.

Kouyeas, V. & Balis, C. (1968). Influence of moisture on the restoration of mycostasis in air-dried soils. *Annls Inst. phytopath. Benaki* 8, 123–144.

Kuhlman, E. G. & Hendrix, F. F. (1964). Infection, growth rate and competitive ability of *Fomes annosus* in inoculated *Pinus echinata* stumps. *Phytopathology* 54, 556–561.

Lai, P. & Bruehl, G. W. (1966). Survival of *Cephalosporium gramineum* in naturally infested wheat straws in soil in the field and in the laboratory. *Phytopathology* 56, 213–218.

Last, F. T. (1960). Longevity of conidia of *Botrytis fabae* Sardiña. *Trans. Br. mycol. Soc.* 43, 673–680.

Lawrence, W. J. C. & Newell, J. (1936). Seedling growth in partially sterilized soil. *Scient. Hort.* 4, 165–177.

Leach, L. D. (1947). Growth rates of host and pathogen as factors determining the severity of pre-emergence damping-off. *J. agric. Res.* 75, 161–179.

Leach, L. D. & Davey, A. E. (1938). Determining the sclerotial population of *Sclerotium rolfsii* by soil analysis and predicting losses of sugar beets on the basis of these analyses. *J. agric. Res.* 56, 619–631.

Leach, R. (1937). Observations on the parasitism and control of *Armillaria mellea. Proc. R. Soc.,* B 121, 561–573.

Leach, R. (1939). Biological control and ecology of *Armillaria mellea* (Vahl) Fr. *Trans. Br. mycol. Soc.* 23, 320–329.

Ledingham, R. J. & Chinn, S. H. F. (1955). A flotation method for obtaining spores of *Helminthosporium sativum* from soil. *Can. J. Bot.* 33, 298–303.

Lester, E. & Shipton, P. J. (1967). A technique for studying inhibition of the parasitic activity of *Ophiobolus graminis* (Sacc.) Sacc. in field soils. *Pl. Path.* 16, 121–123.

Levi, M. P. & Cowling, E. B. (1969). Role of nitrogen in wood deterioration. VII. Physiological adaptation of wood-destroying and other fungi to substrates deficient in nitrogen. *Phytopathology* 59, 460–468.

Levi, M. P., Merrill, W. & Cowling, E. B. (1968). Role of nitrogen in wood deterioration. VI. Mycelial fractions and model nitrogen compounds as substrates for growth of *Polyporus versicolor* and other wood-destroying and wood-inhabiting fungi. *Phytopathology* 58, 626–634.

Linderman, R. G. & Toussoun, T. A. (1968 a). Breakdown in *Thielaviopsis basicola* root rot resistance in cotton by hydrocinnamic (3-phenylpropionic) acid. *Phytopathology* 58, 1431–1432.

Linderman, R. G. & Toussoun, T. A. (1968 b). Predisposition to Thielaviopsis root rot of cotton by phytotoxins from decomposing barley residues. *Phytopathology* 58, 1571–1574.

Lindsey, D. L. (1965). Ecology of plant pathogens in soil. III. Competition between soil fungi. *Phytopathology* 55, 104–110.

Lingappa, B. T. & Lockwood, J. L. (1961). The nature of the widespread soil fungistasis. *J. gen. Microbiol.* 26, 473–485.

Lingappa, B. T. & Lockwood, J. L. (1963). Direct assay of soils for fungistasis. *Phytopathology* **53**, 529–531.

Lingappa, B. T. & Lockwood, J. L. (1964). Activation of soil microflora by fungus spores in relation to soil fungistasis. *J. gen. Microbiol.* **35**, 215–227.

Linskens, H. F. & Haage, P. (1963). Cutinase-Nachweis in phytopathogenen Pilzen. *Phytopath. Z.* **48**, 306–311.

Lloyd, A. B. & Lockwood, J. L. (1966). Lysis of fungal hyphae in soil and its possible relation to autolysis. *Phytopathology* **56**, 595–602.

Lockwood, J. L. (1964). Soil fungistasis. *A. Rev. Phytopath.* **2**, 341–362.

Lucas, R. L. (1955). A comparative study of *Ophiobolus graminis* and *Fusarium culmorum* in saprophytic colonization of wheat straw. *Ann. appl. Biol.* **43**, 134–143.

Mace, M. E. (1965). Isolation and identification of 3-indoleacetic acid from *Fusarium oxysporum* f. *cubense*. *Phytopathology* **55**, 240–241.

Mace, M. E. & Solit, Elinor (1966). Interactions of 3-indoleacetic acid and 3-hydroxytyramine in Fusarium wilt of banana. *Phytopathology* **56**, 245–247.

Macer, R. C. F. (1961 *a*). Saprophytic colonization of wheat straw by *Cercosporella herpotrichoides* Fron and other fungi. *Ann. appl. Biol.* **49**, 152–164.

Macer, R. C. F. (1961 *b*). The survival of *Cercosporella herpotrichoides* Fron in wheat straw. *Ann. appl. Biol.* **49**, 165–172.

Macfarlane, I. (1952). Factors affecting the survival of *Plasmodiophora brassicae* Wor. in the soil and its assessment by a host test. *Ann. appl. Biol.* **39**, 239–256.

Macfarlane, I. (1958). A solution-culture technique for obtaining root-hair or primary infection by *Plasmodiophora brassicae*. *J. gen. Microbiol.* **18**, 720–732.

McKeen, W. E. (1952). *Phialophora radicicola* Cain, a corn rootrot pathogen. *Can. J. Bot.* **30**, 344–347.

McNamara, H. C. & Hooton, D. R. (1933). Sclerotia-forming habits of the cotton root-rot fungus in Texas black-land soils. *J. agric. Res.* **46**, 807–819.

McNamara, H. C., Hooton, D. R. & Porter, D. D. (1931). Cycles of growth in cotton root rot at Greenville, Tex. *Circ. U.S. Dep. Agric.* no. 173.

Marchant, R. & White, Monica F. (1966). Spore swelling and germination in *Fusarium culmorum*. *J. gen. Microbiol.* **42**, 237–244.

Martin, J. P., Aldrich, D. G., Murphy, W. S. & Bradford, G. R. (1953). Effect of soil fumigation on growth and chemical composition of citrus plants. *Soil Sci.* **75**, 137–151.

Martinson, C. A. (1963). Inoculum potential relationships of *Rhizoctonia solani* measured with soil microbiological sampling tubes. *Phytopathology* **53**, 634–638.

Massey, R. E. (1934). *Rep. Gezira agric. Res. Serv. Anglo-Egypt. Sudan, 1933*, pp. 126–146.

Bibliography

Mathew, K. T. (1961). Morphogenesis of mycelial strands in the cultivated mushroom, *Agaricus bisporus*. *Trans. Br. mycol. Soc.* **44**, 285–290.

Maurer, C. L. & Baker, R. (1965). Ecology of plant pathogens in soil. II. Influence of glucose, cellulose and inorganic nitrogen amendments on development of bean root rot. *Phytopathology* **55**, 69–72.

Menzies, J. D. (1959). Occurrence and transfer of a biological factor in soil that suppresses potato scab. *Phytopathology* **49**, 648–652.

Meredith, D. S. (1960). Further observations on fungi inhabiting pine stumps. *Ann. Bot.* **24**, 63–78.

Meronuck, R. A. & Pepper, E. H. (1968). Chlamydospore formation in conidia of *Helminthosporium sativum*. *Phytopathology* **58**, 866–867.

Meynell, G. G. (1957*a*). The applicability of the hypothesis of independent action to fatal infections in mice given *Salmonella typhimurium* by mouth. *J. gen. Microbiol.* **16**, 396–404.

Meynell, G. G. (1957*b*). Inherently low precision of infectivity titrations using a quantal response. *Biometrics* **13**, 149–163.

Meynell, G. G. & Meynell, E. W. (1958). The growth of micro-organisms in vivo with particular reference to the relation between dose and latent period. *J. Hyg. Camb.* **56**, 323–346.

Meynell, G. G. & Stocker, B. A. D. (1957). Some hypotheses on the aetiology of fatal infections in partially resistant hosts and their application to mice challenged with *Salmonella paratyphi-β* or *Salmonella typhimurium* by intraperitoneal injection. *J. gen. Microbiol.* **16**, 38–58.

Millikan, C. R. (1942). Studies on soil conditions in relation to root-rot of cereals. *Proc. R. Soc. Vict.* **54**, 145–195.

Mitchell, R. (1963). Addition of fungal cell-wall components to soil for biological disease control. *Phytopathology* **53**, 1068–1071.

Mitchell, R. & Alexander, M. (1962). Microbiological processes associated with the use of chitin for biological control. *Proc. Soil Sci. Soc. Am.* **26**, 556–558.

Mitchell, R. & Alexander, M. (1963). Lysis of soil fungi by bacteria. *Can. J. Microbiol.* **9**, 169–177.

Mitchell, R. B., Hooton, D. R. & Clark, F. E. (1941). Soil bacteriological studies on the control of Phymatotrichum root rot of cotton. *J. agric. Res.* **63**, 535–547.

Morris, L. G. (1954). The steam sterilizing of soil—experiments on fine soil. *Res. Rep. natn Inst. agric. Engng* no. 14.

Motta, J. (1967). A note on the mitotic apparatus in the rhizomorph meristem of *Armillaria mellea*. *Mycologia* **59**, 370–375.

Moubasher, A. H. (1963). Selective effects of fumigation with carbon disulphide on the soil fungus flora. *Trans. Br. mycol. Soc.* **46**, 338–344.

Mueller, K. E. & Durrell, L. W. (1957). Sampling tubes for soil fungi. *Phytopathology* **47**, 243.

Mughogho, L. K. (1968). The fungus flora of fumigated soils. *Trans. Br. mycol. Soc.* **51**, 441–459.

Müller-Kögler, E. (1938). Untersuchungen über die Schwarzbeinigkeit des Getreides und den Wirtspflanzenkreis ihres Erregers (*Ophiobolus graminis* Sacc.). *Arb. biol. Abt. (Anst.-Reichsanst.) Berl.* **22**, 271–319.

Napper, R. P. N. (1932). A scheme of treatment for the control of *Fomes lignosus* in young rubber areas. *J. Rubb. Res. Inst. Malaya* **4**, 34–38.

Napper, R. P. N. (1934). Root disease investigations. *Rep. Rubb. Res. Inst. Malaya*, 1933, pp. 105–111.

Napper, R. P. N. (1938). Root disease and underground pests in new plantings. *Planter* **19**, 453–455.

Napper, R. P. N. (1940). Root disease investigations. *Rep. Rubb. Res. Inst. Malaya*, 1939, pp. 157–171.

Nash, Shirley M. & Alexander, J. V. (1965). Comparative survival of *Fusarium solani* f. *cucurbitae* and *F. solani* f. *phaseoli* in soil. *Phytopathology* **55**, 963–966.

Nash, Shirley M., Christou, T. & Snyder, W. C. (1961). Existence of *Fusarium solani* f. *phaseoli* as chlamydospores in soil. *Phytopathology* **51**, 308–312.

Nash, Shirley M. & Snyder, W. C. (1962). Quantitative estimations by plate counts of propagules of the bean root rot Fusarium in field soils. *Phytopathology* **52**, 567–572.

Nash Smith, Shirley M. (1969). The significance of populations of pathogenic fusaria in soil. In *Root Diseases and Soil-borne Pathogens*, T. A. Toussoun, R. V. Bega & P. E. Nelson ed. (In Press.) Berkeley: University of California Press.

Nelson, E. E. (1964). Some probable relationships of soil fungi and zone lines to survival of *Poria weirii* in buried wood blocks. *Phytopathology* **54**, 120–121.

Newcombe, Margaret (1960). Some effects of water and anaerobic conditions on *Fusarium oxysporum* f. *cubense* in soil. *Trans. Br. mycol. Soc.* **43**, 51–59.

Noble, R. J. (1924). Studies on the parasitism of *Urocystis tritici*, the organism causing flag smut of wheat. *J. agric. Res.* **27**, 451–489.

Nutman, P. S. (1963). Factors influencing the balance of mutual advantage in legume symbiosis. In *Symbiotic Associations, Thirteenth Symp. Soc. gen. Microbiol.*, P. S. Nutman & Barbara Mosse ed., pp. 51–71. Cambridge University Press.

Papavizas, G. C. (1964). Survival of single-basidiospore isolates of *Rhizoctonia praticola* and *Rhizoctonia solani*. *Can. J. Microbiol.* **10**, 739–746.

Papavizas, G. C., Adams, P. B. & Lewis, J. A. (1968). Survival of root-infecting fungi in soil. V. Saprophytic multiplication of *Fusarium solani* f. sp. *phaseoli* in soil. *Phytopathology* **58**, 414–420.

Papavizas, G. C. & Davey, C. B. (1959). Isolation of *Rhizoctonia solani* Kuehn from naturally infested and artificially inoculated soils. *Pl. Dis. Reptr* **43**, 404–410.

Papavizas, G. C. & Davey, C. B. (1961). Saprophytic behaviour of *Rhizoctonia* in soil. *Phytopathology* **51**, 693–699.

Papavizas, G. C. & Davey, C. B. (1962*a*). Activity of *Rhizoctonia* in soil as affected by carbon dioxide. *Phytopathology* **52**, 759–766.

Papavizas, G. C. & Davey, C. B. (1962*b*). Isolation and pathogenicity of *Rhizoctonia* saprophytically existing in soil. *Phytopathology* **52**, 834–840.

Patrick, Z. A. & Koch, L. W. (1963). The adverse influence of phytotoxic substances from decomposing plant residues on resistance of tobacco to black root rot. *Can. J. Bot.* **41**, 747–758.

Patrick, Z. A. & Toussoun, T. A. (1965). Plant residues and organic amendments in relation to biological control. In *Ecology of Soilborne Plant Pathogens*, K. F. Baker & W. C. Snyder ed., pp. 440–459. Berkeley: University of California Press.

Patrick, Z. A., Toussoun, T. A. & Snyder, W. C. (1963). Phytotoxic substances in arable soils associated with decomposition of plant residues. *Phytopathology* **53**, 152–161.

Peries, O. S., Fernando, T. M. & Samaraweera, S. K. (1963). Field evaluations of methods for the control of white root disease (*Fomes lignosus*) of *Hevea*. *Q. Jl Rubb. Res. Inst. Ceylon* **39**, 9–15.

Perkins, A. J. (1917). Eyre's Peninsula. Its agricultural development and the work of the Department of Agriculture. *J. Dep. Agric. S. Aust.* **20**, 684–690.

Phillips, D. J. (1965). Ecology of plant pathogens in soil. IV. Pathogenicity of macroconidia of *Fusarium roseum* f. sp. *cerealis* produced on media of high or low nutrient content. *Phytopathology* **55**, 328–329.

Presley, J. T. (1939). Unusual features in the behaviour of sclerotia of *Phymatotrichum omnivorum*. *Phytopathology* **29**, 498–502.

Presley, J. T., Carns, H. R., Taylor, E. E. & Schnathorst, W. C. (1966). Movement of conidia of *Verticillium albo-atrum* in cotton plants. *Phytopathology* **56**, 375.

Proctor, P. (1941). Penetration of the walls of wood cells by the hyphae of wood-destroying fungi. *Bull. Univ. Yale Sch. For.* no. 47.

Punter, D. (1963). The effects of stump treatments on fungal colonization of conifer stumps. Ph.D. thesis, University of Cambridge.

Rao, A. S. (1959). A comparative study of competitive saprophytic ability in twelve root-infecting fungi by an agar plate method. *Trans. Br. mycol. Soc.* **42**, 97–111.

Redfern, D. B. (1968). The ecology of *Armillaria mellea* in Britain. Biological control. *Ann. Bot.* **32**, 293–300.

Reinking, O. A. & Manns, M. M. (1933). Parasitic and other fusaria counted in tropical soils. *Z. ParasitKde* **6**, 23–75.

Rifai, M. A. (1969). A revision of the genus *Trichoderma*. *Mycol. Pap.* no. 116.

Rifai, M. A. & Webster, J. (1966*a*). Culture studies on *Hypocrea* and *Trichoderma*. II. *H. aureo-viridis* and *H. rufa* f. *sterilis* f. nov. *Trans. Br. mycol. Soc.* **49**, 289–296.

Rifai, M. A. & Webster, J. (1966*b*). Culture studies on *Hypocrea* and *Trichoderma*. III. *H. lactea* (= *H. citrina*) and *H. pulvinata*. *Trans. Br. mycol. Soc.* **49**, 297–310.

Rishbeth, J. (1950). Observations on the biology of *Fomes annosus*, with particular reference to East Anglian pine plantations. I. The outbreaks of disease and ecological status of the fungus. *Ann. Bot.* **14**, 365–383.

Rishbeth, J. (1951 *a*). Observations on the biology of *Fomes annosus*, with particular reference to East Anglian pine plantations. II. Spore production, stump infection and saprophytic activity in stumps. *Ann. Bot.* **15**, 1–21.

Rishbeth, J. (1951 *b*). Observations on the biology of *Fomes annosus*, with particular reference to East Anglian pine plantations. III. Natural and experimental infection of pines, and some factors affecting severity of the disease. *Ann. Bot.* **15**, 221–246.

Rishbeth, J. (1951 *c*). Butt rot by *Fomes annosus* Fr. in East Anglian conifer plantations and its relation to tree killing. *Forestry* **24**, 114–120.

Rishbeth, J. (1952). Control of *Fomes annosus* Fr. *Forestry* **25**, 41–50.

Rishbeth, J. (1955). Fusarium wilt of bananas in Jamaica. I. Some observations on the epidemiology of the disease. *Ann. Bot.* **19**, 293–328.

Rishbeth, J. (1957). Fusarium wilt of bananas in Jamaica. II. Some aspects of host–parasite relationships. *Ann. Bot.* **21**, 215–245.

Rishbeth, J. (1959 *a*). Stump protection against *Fomes annosus*. I. Treatment with creosote. *Ann. appl. Biol.* **47**, 519–528.

Rishbeth, J. (1959 *b*). Stump protection against *Fomes annosus*. II. Treatment with substances other than creosote. *Ann. appl. Biol.* **47**, 529–541.

Rishbeth, J. (1959 *c*). Dispersal of *Fomes annosus* Fr. and *Peniophora gigantea* (Fr.) Massee. *Trans. Br. mycol. Soc.* **42**, 243–260.

Rishbeth, J. (1960). Factors affecting the incidence of banana wilt ('Panama disease'). *Emp. J. exp. Agric.* **28**, 109–113.

Rishbeth, J. (1963). Stump protection against *Fomes annosus*. III. Inoculation with *Peniophora gigantea*. *Ann. appl. Biol.* **52**, 63–77.

Rishbeth, J. (1964). Stump infection by basidiospores of *Armillaria mellea*. *Trans. Br. mycol. Soc.* **47**, 460.

Rishbeth, J. (1967). Control measures against *Fomes annosus* in Great Britain. *Sect.* 24, *Congr. Int. Un. Forest Res. Org., Munich.*

Rishbeth, J. (1968). The growth rate of *Armillaria mellea*. *Trans. Br. mycol. Soc.* **51**, 575–586.

Rishbeth, J. & Naylor, A. G. (1957). Fusarium wilt of bananas in Jamaica. III. Attempted control. *Ann. Bot.* **21**, 599–609.

Roberts, Florence M. (1943). Factors influencing infection of the tomato by *Verticillium albo-atrum*. *Ann. appl. Biol.* **30**, 327–331.

Robertson, N. F. (1954). Studies on the mycorrhiza of *Pinus sylvestris*. *New Phytol.* **53**, 253–283.

Rogers, C. H. (1936). Apparatus and procedure for separating cotton root rot sclerotia from soil samples. *J. agric. Res.* **52**, 73–79.

Rogers, C. H. (1937). The effect of three- and four-year rotations on cotton root rot in the central Texas Blacklands. *J. Am. Soc. Agron.* **29**, 668–680.

Rogers, C. H. (1942). Cotton root rot studies with special reference to sclerotia, cover crops, rotations, tillage, seeding rates, soil fungicides, and effects on seed quality. *Bull. Tex. agric. Exp. Stn* no. 614.

Rogers, C. H. & Watkins, G. M. (1938). Strand formation in *Phymatotrichum omnivorum*. *Am. J. Bot.* **25**, 244–246.

Rosser, W. R. & Chadburn, Barbara L. (1968). Cereal diseases and their effects on intensive wheat cropping in the East Midland region, 1963–65. *Pl. Path.* **17**, 51–60.

Rovira, A. D. (1965). Plant root exudates and their influence upon soil micro-organisms. In *Ecology of Soil-borne Plant Pathogens*, K. F. Baker & W. C. Snyder ed., pp. 170–186. Berkeley: University of California Press.

Royle, D. J. & Hickman, C. J. (1964*a*). Analysis of factors governing in vitro accumulation of zoospores of *Pythium aphanidermatum* on roots. I. Behaviour of zoospores. *Can. J. Microbiol.* **10**, 151–162.

Royle, D. J. & Hickman, C. J. (1964*b*). Analysis of factors governing in vitro accumulation of zoospores of *Pythium aphanidermatum* on roots. II. Substances causing response. *Can. J. Microbiol.* **10**, 201–219.

Russell, E. J. & Hutchinson, H. B. (1909). The effect of partial sterilization of soil on the production of plant food. *J. agric. Sci., Camb.*, **3**, 111–144.

Sadasivan, T. S. (1939). Succession of fungi decomposing wheat straw in different soils, with special reference to *Fusarium culmorum*. *Ann. appl. Biol.* **26**, 497–508.

Saksena, S. B. (1960). Effect of carbon disulphide fumigation on *Trichoderma viride* and other fungi. *Trans. Br. mycol. Soc.* **43**, 111–116.

Salisbury, E. J. (1942). *The Reproductive Capacity of Plants*. London: Bell.

Samuel, G. (1934). *Rep. Waite agric. Res. Inst.*, 1925–32, p. 26.

Samuel, G. & Garrett, S. D. (1932). *Rhizoctonia solani* on cereals in South Australia. *Phytopathology* **22**, 827–836.

Samuel, G. & Garrett, S. D. (1933). Ascospore discharge in *Ophiobolus graminis* and its probable relation to the development of whiteheads in wheat. *Phytopathology* **23**, 721–728.

Samuel, G. & Garrett, S. D. (1945). The infected root-hair count for estimating the activity of *Plasmodiophora brassicae* Woron. in the soil. *Ann. appl. Biol.* **32**, 96–101.

Sanford, G. B. (1926). Some factors affecting the pathogenicity of *Actinomyces scabies*. *Phytopathology* **16**, 525–547.

Sawada, Y., Nitta, K. & Igarashi, T. (1965). Injury of young plants caused by the decomposition of green manure. Part 2. Causal fungus as an agent of the decomposition. *Soil Pl. Fd, Tokyo* **11**, 241–245.

Scheffer, R. P. (1957). Analysis of Fusarium resistance in tomato by grafting experiments. *Phytopathology* **47**, 328–331.

Scheffer, R. P. & Walker, J. C. (1954). Distribution and nature of Fusarium resistance in the tomato plant. *Phytopathology* **44**, 94–101.

Schreiber, L. R. & Green, R. J. (1962). Comparative survival of mycelium, conidia and microsclerotia of *Verticillium albo-atrum* in mineral soil. *Phytopathology* **52**, 288–289.

Schreiber, L. R. & Green, R. J. (1963). Effect of root exudates on germination of conidia and microsclerotia of *Verticillium albo-atrum* inhibited by the soil fungistatic principle. *Phytopathology* **53**, 260–264.

Schroth, M. N. & Cook, R. J. (1964). Seed exudation and its influence on pre-emergence damping-off of bean. *Phytopathology* **54**, 670–673.

Schroth, M. N. & Hendrix, F. F. (1962). Influence of nonsusceptible plants on the survival of *Fusarium solani* f. *phaseoli* in soil. *Phytopathology* **52**, 906–909.

Schroth, M. N. & Snyder, W. C. (1961). Effect of host exudates on chlamydospore germination of the bean root rot fungus, *Fusarium solani* f. *phaseoli*. *Phytopathology* **51**, 389–393.

Schroth, M. N., Toussoun, T. A. & Snyder, W. C. (1963). Effect of certain constituents of bean exudate on germination of chlamydospores of *Fusarium solani* f. *phaseoli* in soil. *Phytopathology* **53**, 809–812.

Schütte, K. H. (1956). Translocation in the fungi. *New Phytol.* **55**, 164–182.

Scott, P. R. (1967). The biology of the take-all fungus (*Ophiobolus graminis*) under continuous cultivation of winter wheat. Ph.D. thesis, University of Cambridge.

Scott, P. R. (1969). Control of survival of *Ophiobolus graminis* between consecutive crops of winter wheat. *Ann. appl. Biol.* **63**, 37–43.

Sewell, G. W. F. (1959). Direct observation of *Verticillium albo-atrum* in soil. *Trans. Br. mycol. Soc.* **42**, 312–321.

Sewell, G. W. F. & Wilson, J. F. (1964). Occurrence and dispersal of Verticillium conidia in xylem sap of the hop (*Humulus lupulus* L.). *Nature, Lond.* **204**, 901.

Shear, G. M. (1943). Plant tissue tests versus soil tests for determining the availability of nutrients for tobacco. *Tech. Bull. Va agric. Exp. Stn* no. 84.

Shear, G. M. & Wingard, S. A. (1944). Some ways by which nutrition may affect severity of disease in plants. *Phytopathology* **34**, 603–605.

Sholto Douglas, J. (1959). *Hydroponics: the Bengal System*, 3rd ed. Oxford University Press.

Simmonds, P. M., Sallans, B. J. & Ledingham, R. J. (1950). The occurrence of *Helminthosporium sativum* in relation to primary infections in common root rot of wheat. *Scient. Agric.* **30**, 407–417.

Sisler, H. D. & Cox, C. E. (1954). Effects of tetramethylthiuram disulfide on metabolism of *Fusarium roseum*. *Am. J. Bot.* **41**, 338–345.

Smith, F. E. V. (1936). *Rep. Dep. Agric. Jamaica*, 1935, pp. 53–72.

Smith, N. R. & Dawson, Virginia T. (1944). The bacteriostatic action of rose bengal in media used for plate counts of soil fungi. *Soil Sci.* **58**, 467–471.

Smith, Sarah E. (1966). Physiology and ecology of orchid mycorrhizal fungi with reference to seedling nutrition. *New Phytol.* **65**, 488–499.

277

Smith, Sarah E. (1967). Carbohydrate translocation in orchid mycorrhizas. *New Phytol.* **66**, 371–378.

Sneh, B., Katan, J., Henis, Y. & Wahl, I. (1966). Methods for evaluating inoculum density of *Rhizoctonia* in naturally infested soil. *Phytopathology* **56**, 74–78.

Snider, P. J. (1959). Stages of development in rhizomorphic thalli of *Armillaria mellea*. *Mycologia* **51**, 693–707.

Snyder, W. C., Baker, K. F. & Hansen, H. N. (1946). Interpretation of resistance to Fusarium wilt in tomato. *Science, N.Y.* **103**, 707–708.

Snyder, W. C. & Hansen, H. N. (1940). The species concept in *Fusarium*. *Am. J. Bot.* **27**, 64–67.

Snyder, W. C. & Hansen, H. N. (1945). The species concept in *Fusarium* with reference to Discolor and other sections. *Am. J. Bot.* **32**, 657–666.

Snyder, W. C. & Nash, Shirley M. (1968). Relative incidence of Fusarium pathogens of cereals in rotation plots at Rothamsted. *Trans. Br. mycol. Soc.* **51**, 417–425.

Snyder, W. C., Schroth, M. N. & Christou, T. (1959). Effect of plant residues on root rot of bean. *Phytopathology* **49**, 755–756.

Snyder, W. C. & Toussoun, T. A. (1965). Current status of taxonomy in *Fusarium* species and their perfect stages. *Phytopathology* **55**, 833–837.

Spencer, E. L. & McNew, G. L. (1938). The influence of mineral nutrition on the reaction of sweet corn seedlings to *Phytomonas stewarti*. *Phytopathology* **28**, 213–223.

Sprague. R. (1931). *Cercosporella herpotrichoides*, the cause of the Columbia Basin footrot of winter wheat. *Science, N.Y.* **74**, 51–53.

Stover, R. H. (1950). The black root rot disease of tobacco. I. Studies on the causal organism, *Thielaviopsis basicola*. *Can. J. Res., C*, **28**, 445–470.

Stover, R. H. (1962). Studies on Fusarium wilt of bananas. IX. Competitive saprophytic ability of *Fusarium oxysporum* f. *cubense*. *Can. J. Bot.* **40**, 1473–1481.

Stover, R. H. (1969). Banana root diseases caused by *Fusarium oxysporum* f. sp. *cubense, Pseudomonas solanacearum,* and *Radopholus similis*: a comparative study of life cycles in relation to control. In *Root Diseases and Soil-borne Pathogens*, T. A. Toussoun, R. V. Bega & P. E. Nelson ed. (In Press.) Berkeley: University of California Press.

Stover, R. H. & Waite, B. H. (1960). Studies on Fusarium wilt of bananas. V. Pathogenicity and distribution of *F. oxysporum* f. *cubense* races 1 and 2. *Can. J. Bot.* **38**, 51–61.

Struckmeyer, B. Esther, Beckman, C. H., Kuntz, J. E. & Riker, A. J. (1954). Plugging of vessels by tyloses and gums in wilting oaks. *Phytopathology* **44**, 148–153.

Stumbo, C. R., Gainey, P. L. & Clark, F. E. (1942). Microbiological and nutritional factors in the take-all disease of wheat. *J. agric. Res.* **64**, 653–665.

Sussman, A. S. (1966). Types of dormancy as represented by conidia and ascospores of *Neurospora*. In *The Fungus Spore*, Colston Papers no. 18, M. F. Madelin ed., pp. 235–257. London: Butterworth.

Swift, M. J. (1968). Inhibition of rhizomorph development by *Armillaria mellea* in Rhodesian forest soils. *Trans. Br. mycol. Soc.* **51**, 241–247.

Talboys, P. W. (1958a). Some mechanisms contributing to Verticillium-resistance in the hop root. *Trans. Br. mycol. Soc.* **41**, 227–241.

Talboys, P. W. (1958b). Association of tylosis and hyperplasia of the xylem with vascular invasion of the hop by *Verticillium albo-atrum*. *Trans. Br. mycol. Soc.* **41**, 249–260.

Talboys, P. W. (1958c). Degradation of cellulose by *Verticillium albo-atrum*. *Trans. Br. mycol. Soc.* **41**, 242–248.

Talboys, P. W. (1964). A concept of the host–parasite relationship in Verticillium wilt diseases. *Nature, Lond.* **202**, 361–364.

Tam, R. K. & Clark, H. E. (1943). Effect of chloropicrin and other soil disinfectants on the nitrogen nutrition of the pineapple plant. *Soil Sci.* **56**, 245–261.

Taubenhaus, J. J. & Ezekiel, W. N. (1931). Cotton root rot and its control. *Bull. Tex. agric. Exp. Stn* no. 423.

Thornton, R. H. (1952). The screened immersion plate. A method of isolating soil micro-organisms. *Research, Lond.* **5**, 190–191.

Thornton, R. H. (1953). Features of growth of *Actinomyces* in soil. *Research, Lond.* **6**, no. 6.

Toussoun, T. A., Nash, Shirley M. & Snyder, W. C. (1960). The effect of nitrogen sources and glucose on the pathogenesis of *Fusarium solani* f. *phaseoli*. *Phytopathology* **50**, 137–140.

Toussoun, T. A. & Nelson, P. E. (1968). *A Pictorial Guide to the Identification of Fusarium species according to the Taxonomic System of Snyder & Hansen*. Pennsylvania State University Press.

Toussoun, T. A. & Patrick, Z. A. (1963). Effect of phytotoxic substances from decomposing plant residues on root rot of bean. *Phytopathology* **53**, 265–270.

Toussoun, T. A., Patrick, Z. A. & Snyder, W. C. (1963). Influence of crop residue decomposition products on the germination of *Fusarium solani* f. *phaseoli* chlamydospores in soil. *Nature, Lond.* **197**, 1314–1316.

Toussoun, T. A. & Snyder, W. C. (1961). Germination of chlamydospores of *Fusarium solani* f. *phaseoli* in unsterilized soils. *Phytopathology* **51**, 620–623.

Toussoun, T. A., Weinhold, A. R., Linderman, R. G. & Patrick, Z. A. (1968). Nature of phytotoxic substances produced during plant residue decomposition in soil. *Phytopathology* **58**, 41–45.

Towers, B. & Stambaugh, W. J. (1968). The influence of induced soil moisture stress upon *Fomes annosus* root rot of loblolly pine. *Phytopathology* **58**, 269–272.

Tribe, H. T. (1960). Decomposition of buried cellulose film, with special reference to the ecology of certain soil fungi. In *The Ecology of*

Soil Fungi, D. Parkinson & J. S. Waid ed., pp. 246–256. Liverpool University Press.

Tribe, H. T. (1966). Interactions of soil fungi on cellulose film. *Trans. Br. mycol. Soc.* **49**, 457–466.

Trujillo, E. E. (1963). Pathological-anatomical studies of Gros Michel banana affected by Fusarium wilt. *Phytopathology* **53**, 162–166.

Trujillo, E. E. & Hine, R. B. (1965). The role of papaya residues in papaya root rot caused by *Pythium aphanidermatum* and *Phytophthora parasitica*. *Phytopathology* **55**, 1293–1298.

Trujillo, E. E. & Snyder, W. C. (1963). Uneven distribution of *Fusarium oxysporum* f. *cubense* in Honduras soils. *Phytopathology* **53**, 167–170.

Turner, P. D. (1965). The incidence of Ganoderma disease of oil palms in Malaya and its relation to previous crop. *Ann. appl. Biol.* **55**, 417–423.

Turner, P. D. & Bull, R. A. (1967). *Diseases and Disorders of the Oil Palm in Malaysia*. Kuala Lumpur: Incorporated Society of Planters.

Tyner, L. E. (1966). Associative effects of fungi on *Cochliobolus sativus*. *Phytopathology* **56**, 776–780.

Valder, P. G. (1958). The biology of *Helicobasidium purpureum* Pat. *Trans. Br. mycol. Soc.* **41**, 283–308.

Van der Plank, J. E. (1947). The relation between the size of the plant and the spread of systemic diseases. I. A discussion of ideal cases and a new approach to problems of control. *Ann. appl. Biol.* **34**, 376–387.

Vanterpool, T. C. (1940). Present knowledge of browning root rot of wheat with special reference to its control. *Scient. Agric.* **20**, 735–749.

Vanterpool, T. C. (1952). The phenomenal decline of browning root rot (*Pythium* spp.) on the Canadian Prairies. *Scient. Agric.* **32**, 443–452.

Waksman, S. A. (1932). *Principles of Soil Microbiology*, 2nd ed. London: Baillière, Tindall and Cox.

Walker, A. G. (1941). The colonization of buried wheat straw by soil fungi, with special reference to *Fusarium culmorum*. *Ann. appl. Biol.* **28**, 333–350.

Wallis, G. W. (1961). Infection of Scots pine roots by *Fomes annosus*. *Can. J. Bot.* **39**, 109–121.

Wallis, G. W. & Reynolds, G. (1965). The initiation and spread of *Poria weirii* root rot of Douglas fir. *Can. J. Bot.* **43**, 1–9.

Warcup, J. H. (1950). The soil-plate method for isolation of fungi from soil. *Nature, Lond.* **166**, 117.

Warcup, J. H. (1955). On the origin of colonies of fungi developing on soil dilution plates. *Trans. Br. mycol. Soc.* **38**, 298–301.

Wardlaw, C. W. (1941). The banana in Central America. III. Panama disease. *Nature, Lond.* **147**, 380–381.

Wastie, R. L. (1961). Factors affecting competitive saprophytic colonization of the agar plate by various root-infecting fungi. *Trans. Br. mycol. Soc.* **44**, 145–159.

Wastie, R. L. (1962). Mechanism of action of an infective dose of *Botrytis* spores on bean leaves. *Trans. Br. mycol. Soc.* **45**, 465–473.

Waterston, J. M. (1941). Observations on the parasitism of *Rosellinia pepo* Pat. *Trop. Agric., Trin.*, **18**, 174–184.

Webb, P. C. R. (1949). Zoosporangia, believed to be those of *Plasmodiophora brassicae*, in the root hairs of non-cruciferous plants. *Nature, Lond.* **163**, 608.

Webster, J. (1964). Culture studies on *Hypocrea* and *Trichoderma*. I. Comparison of perfect and imperfect states of *H. gelatinosa, H. rufa* and *Hypocrea* sp. 1. *Trans. Br. mycol. Soc.* **47**, 75–96.

Webster, J. & Lomas, Norma (1964). Does *Trichoderma viride* produce gliotoxin and viridin? *Trans. Br. mycol. Soc.* **47**, 535–540.

Webster, J. & Rifai, M. A. (1968). Culture studies on *Hypocrea* and *Trichoderma*. IV. *Hypocrea pilulifera* sp. nov. *Trans. Br. mycol. Soc.* **51**, 511–514.

Weinhold, A. R. & Garraway, M. O. (1966). Nitrogen and carbon nutrition of *Armillaria mellea* in relation to growth-promoting effects of ethanol. *Phytopathology* **56**, 108–112.

White, N. H. (1947). The etiology of take-all disease of wheat. III. Factors concerned with the development of take-all symptoms in wheat. *J. Coun. scient. ind. Res. Aust.* **20**, 66–81.

White, N. H. (1954). The use of decoy crops in the eradication of certain soil-borne plant diseases. *Aust. J. Sci.* **17**, 18–19.

Whitney, N. J. (1954). Investigations on *Rhizoctonia crocorum* (Pers.) DC. in relation to the violet root rot of carrot. *Can. J. Bot.* **32**, 679–704.

Wiehe, P. O. (1952). The spread of *Armillaria mellea* (Fr.) Quél. in tung orchards. *E. Afr. agric. J.* **18**, 67–72.

Wilhelm, S. (1965). *Pythium ultimum* and the soil fumigation growth response. *Phytopathology* **55**, 1016–1020.

Wilhelm, S. & Nelson, P. E. (1969). A concept of rootlet health of strawberries in pathogen-free field soil achieved by fumigation. In *Root Diseases and Soil-borne Pathogens*, T. A. Toussoun, R. V. Bega & P. E. Nelson ed. (In Press.) Berkeley: University of California Press.

Wilhelm, S., Storkan, R. C. & Sagen, J. E. (1961). Verticillium wilt of strawberry controlled by fumigation of soil with chloropicrin and chloropicrin-methyl bromide mixtures. *Phytopathology* **51**, 744–748.

Wilhelm, S., Storkan, R. C., Sagen, J. E. & Carpenter, T. (1959). Large-scale soil fumigation against broomrape. *Phytopathology* **49**, 530–531.

Williams, P. G., Scott, K. J. & Kuhl, Joy L. (1966). Vegetative growth of *Puccinia graminis* f. sp. *tritici* in vitro. *Phytopathology* **56**, 1418–1419.

Wilson, A. R. (1937). The chocolate spot disease of beans (*Vicia faba* L.) caused by *Botrytis cinerea* Pers. *Ann. appl. Biol.* **24**, 258–288.

Wood, R. K. S. (1967). *Physiological Plant Pathology*. Oxford: Blackwell.

Woodruff, H. B. (1966). The physiology of antibiotic production: the role of the producing system. In *Biochemical Studies of Antimicrobial Drugs*, Sixteenth Symp. Soc. Gen. Microbiol., B. A. Newton & P. E. Reynolds ed., pp. 22–46. Cambridge University Press.

Woronin, M. S. (1878). *Plasmodiophora brassicae*, Urheber der Kohlpflanzen-Hernie. *Jb. wiss. Bot.* **11**, 548–574.

Zarka, A. M. El (1963). A rapid method for the isolation and detection of *Rhizoctonia solani* Kühn from naturally infested and artificially inoculated soils. *Meded. LandbHoogesch. OpzoekStns Gent* **28**, 877–885.

Zaroogian, G. E. & Beckman, C. H. (1968). A comparison of cell wall composition in banana plants resistant or susceptible to *Fusarium oxysporum* f. sp. *cubense. Phytopathology* **58**, 733–735.

Zeevaart, J. A. D. (1955). Aminozaren als koolstofbron voor *Fusarium oxysporum* in de houtvaten van lupine planten. *Tijdschr. PlZiekt.* **61**, 76–78.

Zentmyer, G. A. (1961). Chemotaxis of zoospores for root exudates. *Science, N.Y.* **133**, 1595–1596.

Index

Aerated steam, for partial sterilization of soil, 237–8

aeration of soil
control by soil cultivation, 235
affecting *Fusarium* spp., production of macroconidia and chlamydospores by, 192–3
Ophiobolus graminis, spread of, 106–7; survival of, 147–9
Pythium ultimum, seedling killing by, 37
Rhizoctonia solani, mycelial growth of, 138

Agaricus bisporus, morphogenesis of mycelial strands, 96

air-borne pathogenic fungi
colonizing plant shoot-systems, 112–13, 127–8
compared with soil-borne pathogens, 1–2, 207–8

ammonia phytotoxicity after soil steaming, 238

ammonium nitrogen, preference of some plant species for, 239–40

anastomosis
between hyphae in strand formation, 95, 98–9
between mycelial strands, 94

animals, infectivity titrations against, 15

antibiotics
antibacterial, reducing soil fungistasis, 195
nutritional conditions for production, 161–2
onion roots producing, 221–2
persistence in soils, 108
production by soil fungi, 115–16, 123, 161–2
tolerance by soil fungi, 115–16, 123–5, 131–5, 221–2
Trichoderma spp. producing, 108, 169–73 *passim*
use in isolation media, 130, 137, 139

Aphanomyces euteiches, infecting pea seedlings, 37

apparent spread of vascular wilt diseases, 34

apple replant disease, 53

appressoria of *Rhizoctonia solani*, 41–2

arboricides, use in root-disease control, 9, 248–9, 253–4

Armillaria mellea
authority for taxonomic binomial, 7
basidiospores, infectivity, 252
declining resistance of senescent roots, 7, 86, 253–4
potato tubers infected, 10, 92
pseudosclerotium and zone-lines, 167–9
rhizomorphs, morphogenesis and behaviour, 9–10, **99–106**, 253–4
ring-barking as control method, 8, 141, 248, 253–4
saprophytic in coal mines, 142
soil fumigation controlling, 169–73
soil moisture content for rhizomorph growth, 105–6
soil toxin repressing rhizomorph production, 105
substrate value of living *v* dead host tissues, 141–2
temperature, affecting rhizomorph production, 102–5; root infection, 33
Trichoderma spp. antagonistic towards, 169–73

autolysis of mycelium, 197–8, 212

avocado, root infection by *Phytophthora cinnamomi*, 36

Baiting soil for isolation of fungi, 110–11

banana
growth habit of root system, 71–2
Panama disease, *see Fusarium oxysporum* f. *cubense*

bean (*Phaseolus vulgaris*)
Fusarium foot-rot, *see Fusarium solani* f. *phaseoli*
pre-emergence killing by fungi, 37

bean (*Vicia faba*), synergism in aggressive phase of chocolate spot, 88–9

biological control of root diseases, 9, **28–30**, 50, 169, 212, 233–4, 245

Botrytis cinerea, germination and infectivity of spores, 11–12, 16, 181

Merulius lacrymans, morphogenesis of mycelial strands, 96–8
Musa balbisiana, for detection of *Fusarium oxysporum* f. *cubense*, 194
mycelial aggregation into sheets, strands and rhizomorphs, 92–106
mycelial growth rate
  compared with rate of tissue penetration, 123–5
  inverse correlation with specialization of parasitism, 23, 35, 135
  of primary saprophytic sugar fungi, 115–16, 133–4, 137
mycelial growth through soil, 43–4
mycelial strands, 94–9
mycorrhizas, of forest trees, 1–2, 24, 80–2; of orchid seedlings, 45–6

Nematodes, in aetiology of peach replant disease, 54
nitrogen available in soil
  affecting longevity of cereal pathogens in wheat straw, 151–60
  affecting take-all of cereals, 6, 51, 232
  affecting vascular wilts, 70–9 *passim*
  changes in, following partial sterilization of soil, 239–40
  controlling cellulolysis rate, 149–51
  translocated in organic form up xylem, 77
nitrogen, fungal utilization in saprophytic competition, 129; economy of, by wood-rotting fungi, 163–4
nutrient agar, competitive saprophytic colonization by fungi, 129–36
nutrient content of fungal spores, 175–82 *passim*
nutrient deficiencies in soil, predisposing plants to disease
  minor elements, due to over-liming, 235
  nitrogen, aggravating take-all of cereals, 6, 51
  phosphate, causing browning root-rot of cereals, 49; aggravating take-all of cereals, 6
  potash, aggravating Panama disease of bananas, 75–6
  zinc, causing cereal root-rot, 50

nutrient exudation
  by roots, *see* root exudates
  by translocating fungal hyphae, 97–8

Oak wilt due to *Ceratocystis fagacearum*, 66–7
obligate parasites, 24–6
oil palm, basal stem rot due to *Ganoderma* spp., 254–5
oleoresin, *see* resin
onion, white rot disease, 219–22
oospores of *Phytophthora cactorum*, 183–5
Ophiobolus graminis
  ascospores, infection by, 16–18, 186–8
  authority for taxonomic binomial, 6
  biological control, 228
  carbon dioxide affecting ectotrophic infection, 107
  causing take-all on 'wrong' soil type, 21
  competitive saprophytic colonization by, 120–33 *passim*
  control, by crop rotation, 160, 226; direct drilling of wheat, 231; firm seed-bed, 235; rotavation of infected stubble, 232; undersown catch-crop, 232
  decline of take-all under wheat monoculture, 227–9
  diagnosis of infection by, 147, 158
  ectotrophic infection-habit, 82–4, 106–7
  escape of wheat plants from take-all, 5–6, 227–9
  host range, 28, 226
  introduction into Zuider Zee polders, 186–8, 228
  saprophytic survival, 145–61 *passim*
  soil conditions affecting infection by, 5–6, 21, 106–7, 230
  straw tissue penetration rate, 123–5
  surface sterilants for isolation of, 136
  temperature affecting saprophytic competition by, 127
  tolerance of competition, 124–5, 132
  wheat seedling test for detection of, 121, 146
orchid seedlings, mycorrhizal nutrition, 45–6